SpringerBriefs in Applied Sciences and Technology

For further volumes:
http://www.springer.com/series/8884

Wolfgang Rainer Fahrner
Editor

Amorphous Silicon / Crystalline Silicon Heterojunction Solar Cells

 Chemical Industry Press

Springer

Editor
Wolfgang Rainer Fahrner
University of Hagen
Hagen
Germany

and

School of Photovoltaic Engineering
Nanchang University
Nanchang
China

ISSN 2191-530X   ISSN 2191-5318   (electronic)
ISBN 978-3-642-37038-0   ISBN 978-3-642-37039-7   (eBook)
DOI 10.1007/978-3-642-37039-7
Springer Heidelberg New York Dordrecht London

Jointly published with Chemical Industry Press, Beijing
ISBN: 978-7-122-15935-9 Chemical Industry Press, Beijing

Library of Congress Control Number: 2013934387

Printed on acid-free paper

Springer is part of Springer Science+Business Media (www.springer.com)

Preface

The need to replace conventional energies—coal, oil and nuclear power—by alternative ones has been emphasised many times and underlined only recently in the Durban Climate Change Conference. Among these alternatives, photovoltaic devices play a leading role. This book here deals with one important representative, the heterojunction solar cell.

As its name points out, it consists of two different materials, crystalline and amorphous silicon. While the former one was brought to a high standard already shortly after World War II, amorphous silicon was investigated in detail only in 1968, in Romania. In contrast, heterojunction solar cell production of today is a flourishing business as seen by the example of Sanyo or Meyer Burger.

This book deals with some typical properties of the heterojunction cell. Its history, schematic cross-sections, and production tools will be shown. A special chapter is devoted to the challenges of the cell such as texturization, interface defects, passivation, lifetime and surface velocity, epitaxial layer formation, emitter, and back surface field conductivity.

Some important measurement tools are presented.

Today no electronic device will be produced any more before it is not simulated. Thus, we present a few of the simulation programmes available on the market.

The book is completed with a brief survey of the state of the art as represented by the efficiencies.

Because China is the strongest emerging market in the solar cell field a collection of related publications and their discussion appeared to be mandatory.

W. R. Fahrner

Nanchang, December 2011

Wolfgang Rainer Fahrner

Acknowledgments

The authors thank Mrs. K. Meusinger and Dipl.-Ing. B. Wdowiak, University of Hagen, for technical assistance, A. Denker, affiliated to the Helmholtz Zentrum Berlin (HZB) (formerly Hahn-Meitner Institut Berlin (HMI)) and B. Limata, M. Romano and L. Gialanella from the Physics Department of Naples University for the proton irradiation, R. Scheer (HZB) for the EBIC measurements and M. Ferrara affiliated to the ENEA research center in Portici for the electroluminescence measurements. The authors gratefully acknowledge the fruitful discussions with M. Kunst (HZB).

Contents

Contributors

Wolfgang Rainer Fahrner Chair of Electronic Devices, University of Hagen, Haldener Str. 182, 58084 Hagen, Germany; School of Photovoltaic Engineering, Nanchang University, Xuefu Ave. 999, 330031 Nanchang, China, e-mail: wolfgang.fahrner@fernuni-hagen.de

Haibin Huang School of Photovoltaic Engineering, Nanchang University, Xuefu Ave. 999, 330031 Nanchang, China, e-mail: Haibinhuang@ncu.edu.cn

Thomas Mueller Chair of Electronic Devices, University of Hagen, Haldener Str. 182, 58084 Hagen, Germany, e-mail: thomas.mueller@nus.edu.sg

Stefan Schwertheim Chair of Electronic Devices, University of Hagen, Haldener Str. 182, 58084 Hagen, Germany, e-mail: s.schwertheim@web.de

Frank Wuensch Chair of Electronic Devices, University of Hagen, Haldener Str. 182, 58084 Hagen, Germany, e-mail: Frank.Wuensch@alumni.TU-Berlin.de

Heinz-Christoph Neitzert University of Salerno, Via Ponte Don Melillo 1, 84084 Fisciano, SA, Italy, e-mail: neitzert@unisa.it

Amorphous Silicon / Crystalline Silicon Heterojunction Solar Cells

Wolfgang Rainer Fahrner

1 Introduction

1.1 Basic Structure

Like any other (semiconductor) solar cell, the amorphous silicon / crystalline silicon heterojunction solar cell consists of a combination of p-type and n-type material, that is, a diode structure. However, while in the usual case the n-type and the p-type semiconductors are identical and just differ in the doping, a heterojunction is built on two different materials, crystalline and amorphous silicon in our case. Its basic structure is given in Fig. 1.

A crystalline wafer acts as a substrate for the amorphous layer on its top. In the following, the abbreviations c-Si and a-Si are used for crystalline and amorphous silicon, respectively. To obtain the required diode structure, it is evident that the two materials must be of opposite doping type. The amorphous layer acts as emitter and the wafer as the base of the solar cell.

During the development of the heterojunction cell, the range of emitter and base materials has been expanded. For instance, the initially monocrystalline silicon had been replaced by multicrystalline material of various origins (edge-defined ribbon growth, block-cast silicon, etc.). Similarly, microcrystalline silicon had been tried instead of amorphous silicon.

Of course, a cell according to Fig. 1 would not operate very efficiently. For instance, incoming light would be reflected to a good deal. As a first counter-measure, an antireflection coating (ARC, cf. Sect. 3.2) is deposited on the a-Si. The ARC is highly transparent and highly conductive. Together with a texture

W. R. Fahrner (✉)
Chair of Electronic Devices, University of Hagen, Haldener 182, 58084 Hagen, Germany
e-mail: wolfgang.fahrner@fernuni-hagen.de

W. R. Fahrner
School of Photovoltaic Engineering, Nanchang University, Xuefu Ave. 999, 330031
Nanchang, China

W. R. Fahrner, *Amorphous Silicon / Crystalline Silicon Heterojunction Solar Cells,*
SpringerBriefs in Applied Sciences and Technology, DOI: 10.1007/978-3-642-37039-7_1,
© Chemical Industry Press, Beijing and Springer-Verlag Berlin Heidelberg 2013

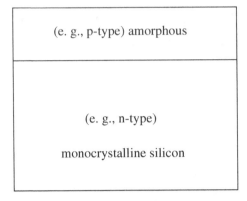

Fig. 1 Basic structure of the amorphous silicon / crystalline silicon heterojunction solar cell

(roughening of the surface, cf. Sect. 3.2), reflection is almost totally suppressed. Furthermore, we need electrical contacts. The usual solution is an array of metallic fingers mutually connected by a busbar on the front side and a metal back contact. However, other solutions with interdigitated contacts at the rear are known (cf. Sect. 4.2). The application of a conductive ARC serves a second purpose, namely the improved lateral transport of carriers generated between the fingers. The a-Si alone offers a very low conductance. Finally, the surfaces of the wafer must not be neglected. The c-Si surfaces are full of so-called surface (or interface) states yielding a high recombination and loss in current gain. Thus, it is advisable to deposit a thin intrinsic layer on either surface of the wafer (cf. Sect. 5.3). As a rule, the deposition of the backside intrinsic layer is followed by another deposition, namely the so-called back surface field (BSF) layer. The BSF repels minority carriers trying to reach the recombination centers at the back contact.

Summing up all these improvements, we end up with a cross section of Fig. 2.

(a)

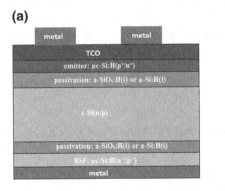

(b)

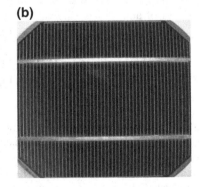

Fig. 2 a Improved structure (*left*), texture not shown and **b** a photograph of a typical solar cell (*right*)

Typical geometrical data of the solar cell are Area of 100×100 mm, wafer thickness of 210 μm, finger width of 150 μm, finger distance (center to center) of 2 mm, busbar width of 2 mm, busbar distance (center to center) of 50 mm, intrinsic layer thicknesses of 5–10 nm, indium tin oxide layer (as ARC) of 80 nm, emitter layer of 5–10 nm, and the metallization on the back side of a few μm.

1.2 History of a-Si:H/c-Si Device Development

The investigation of the heterojunction between amorphous silicon and crystalline silicon started more than 40 years ago. The first-reported a-Si/c-Si heterojunction has been published by Grigorivici et al. [1], by evaporation of amorphous non-hydrogenated silicon at room temperature on top of n-type or p-type crystalline silicon with different resistivities and successive annealing of the deposited films at temperatures of either 290 or 500 °C [1]. This deposition method resulted in a highly defective amorphous silicon layer, and the study aimed to determine the transport mechanism in the amorphous silicon layer. The authors concluded that the a-Si had p-type character from the better rectification, observed in the case of the a-Si/n-type c-Si heterojunction as compared to the a-Si/p-type c-Si heterojunction. The same results had some years earlier found for amorphous germanium on crystalline silicon heterojunctions [2]. From capacitance–voltage (C–V) characteristics of this heterojunctions, the authors deduced an acceptor density in the a-Si of 10^{16} cm^{-3}. Only few years later, good rectification has been obtained on a similar structure and, using capacitance–voltage measurements for the electrical characterization, a rather high built-in-voltage of 0.7 V has been measured [3]. The authors attributed the improved characteristics to a better interface, obtained by heating under vacuum conditions of the crystalline silicon substrate to 400 °C prior of the amorphous silicon deposition. More than 20 years later, in situ transient microwave conductance (TRMC) measurements during heating and cooling of crystalline silicon wafers confirmed the importance of the c-Si substrate heating before amorphous silicon deposition, measuring directly the change in the minority carrier lifetime during this heat treatment [4]. In 1974, the first deposition of hydrogenated amorphous silicon on crystalline silicon has been reported by Fuhs et al. [5], resulting in an amorphous silicon top layer with a lower defect density, where electrical transport in the amorphous layer was not anymore dominated by variable-range hopping [6]. It should be noted that a series of other types of amorphous semiconductor on crystalline silicon heterojunctions have been investigated in these early years, for example, with amorphous germanium [2], amorphous oxides [7], and amorphous chalcogenides [7] as top layers. Photovoltaic effects observed on these devices are already mentioned in [8]. A detailed analysis of the current transport mechanisms in intrinsic amorphous silicon on crystalline silicon heterojunction and a detailed comparison with Schottky diodes have been given for the first time by Brodsky et al. [8]. The authors discussed already the difference of the barrier heights depending on the substrate doping polarity and found a higher barrier for the a-Si/p-type c-Si junction. In the beginning, all a-Si/c-Si heterojunctions have been realized using intrinsic amorphous silicon either on n-type or on p-type Czochralski grown crystalline substrate. Only in 1975, the possibility of a substitutional doping of amorphous silicon has been demonstrated [9]. In the following years, the actually still-used multitunneling capture-emission model, regarding the electronic transport over at the heterojunction in an i-a-Si:H on p-c-Si with different c-Si resistivities, has been developed by Matsuura et al. [10], based on C–V and temperature-dependent dark I–V measurements. Often, as already mentioned for the first publications, the heterojunction has been just used to determine the material parameters of the amorphous silicon top layers, as for example using C–V measurements, an intrinsic a-Si:H on n-type crystalline silicon structure by Sasaki et al. [11]. Similar measurements, but in a MOS configuration, have been performed later, where the highly doped c-Si substrate served just as gate electrode and for the growth of a high-quality thermal oxide [12]. First industrial applications of the heterojunction focused on the vidicon operation and in this case, indeed, a top intrinsic amorphous silicon layer is required in order to avoid lateral current transport and hence "blooming" effects [13].

Some years later, however, a strong interest in photovoltaic applications of the a-Si:H/c-Si heterojunction emerged. Initially, n-type a-Si:H on p-type polycrystalline silicon as low-cost double-junction cells with a nip a-Si:H top cell [14] (cf. Fig. 3) or a-Si:H/ribbon c-Si as bottom cell of a TANDEM solar cell with a top pin-a-Si:H junction [15] has been realized. In the case of the TANDEM device, a three-terminal configuration has been chosen in order to avoid the top/bottom cell current matching problem and an initial efficiency of 11 % has been reported.

In this early period, all photovoltaic a-Si/c-Si heterojunction devices are n-type amorphous on p-type (poly)crystalline devices. As mentioned earlier, the importance of the interface preparation for this kind of heterojunction was already taken into account in a very early stage [3]. More emphasis on the interface recombination influence on the device properties has been given in the moment, when specific characterization tools have been developed. In particular, the application of the contactless (TRMC) technique [16] as in situ technique first for the characterization of a-Si:H layer growth on glass substrates [17, 18] and then on crystalline silicon substrates [19] enabled the monitoring of the interface recombination during the growth of the amorphous layer. In the case of the monitoring of the heterojunction formation, initially a strong damaging of the crystalline silicon substrate due to the plasma process, followed by a successive passivation during a-Si:H growth, has been observed [19, 20]. The combination of the in situ TRMC technique with in situ spectrally resolved ellipsometry enabled, at the same time, a monitoring of the electronic properties and of the structural evolution of the c-Si surface during plasma processing [21]. Another result of the in situ TRMC-measurements is the direct observation of the charge carrier injection from the amorphous layer into the crystalline silicon substrate. In particular, the dependence

Fig. 3 **a** Cross-sectional
view of the tandem solar cell
by Okuda, Okamoto, and
Hamakawa. **b** Schematic
band diagram [14]. Copyright
1983 The Japan Society of
Applied Physics

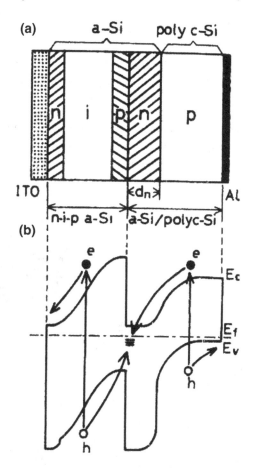

of the injection efficiency on the a-Si:H bulk layer and interface properties has
been investigated [22] and correlated with the electrical characteristics of the
finished heterojunction device [23]. Due to the knowledge that amorphous silicon
layers are as efficient as SiO_2 layers for the surface passivation of crystalline
silicon substrates, as determined again by the TRMC technique [24]. In 1992, a
new type of a-Si:H/c-Si heterojunction solar cell has been presented by Sanyo—
namely the HIT solar cell, the p-type amorphous on n-type crystalline silicon solar
cell with a thin intrinsic interface layer [25]. Already, the first publications
reported 18.1 % conversion efficiency (30), and nowadays, 23.0 % have been
achieved for a large-area (100 cm^2) HIT solar cell [26] (note that Sanyo uses the
acronym HIT, "Heterojunction with Intrinsic Thin layer," for its heterojunction
solar cells).

1.3 Economic Aspects

It might be helpful to start this chapter with a brief survey on the photovoltaic installations. Figure 4 shows the repartition of the power production for Germany, the rest of Europe, and the rest of the world [27].

A more detailed picture is obtained when looking at the annually installed power. An example is found in Fig. 5 for the year 2008 [28].

The interesting point in this figure is the absolute numbers in some segments. Spain, for example, has installed a capacity of 2,400 MW which is more than the

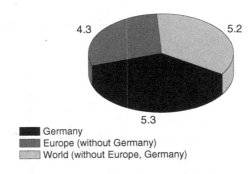

■ Germany
■ Europe (without Germany)
▨ World (without Europe, Germany)

Fig. 4 Repartition of the total installed solar power (in GW) up to 2009

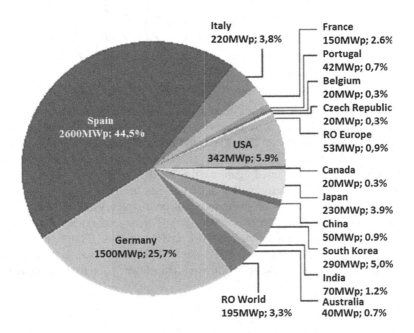

Fig. 5 Worldwide installed solar power in 2008

Table 1 Technical data of a typical Sanyo module [31]

Maximum power (P_{max})	225 Wp*
Voltage at maximum power point (V_{mp})	33.9 V
Current at maximum power point (I_{mp})	6.64 A
Open-circuit voltage (V_{oc})	41.8 V
Short-circuit current (I_{sc})	7.14 A
Power tolerance	+10/−5 %
Maximum system voltage	1,000 V
Dimensions (L × W × H, mm)	1,610 × 861 × 35 mm
Weight	16.5 kg
Performance guarantee	20 years for 80 % of the minimum power
Product guarantee	2 years

output of the Three Miles Island Reactor with an output of 800 and 900 MW of both blocks. It is anticipated that the 2020 installation will amount to 44 GW resulting in accumulated power of 269 GW.

This situation is reflected in the field of heterojunction solar cells. Because Sanyo Company is the only commercial supplier, it is sufficient to use its production data. While in 2009, 340 MW are anticipated, this number will increase to 2 GW in 2020 [29, 30].

Finally, we reproduce some technical data of a typical heterojunction module (Table 1).

It is clearly seen that its minimum efficiency of 16.2 % is one of the best of all commercially available modules. A further advantage consists in the fact that heterojunction modules exhibit a lower open-circuit voltage temperature coefficient (<−1.78 mV/K) than crystalline modules (−0.2 mV/K) [32, 33]. This coefficient essentially determines the temperature coefficient of the efficiency. Recently, Sanyo has repeated these measurements and given more precise data. Temperature coefficients of −0.5 and −0.3 %/K for crystalline and HIT solar cells, respectively, were given [34].

2 Useful Material Parameters

2.1 Useful Data of Monocrystalline Silicon

The crystal structure of the most frequently used semiconductor material, namely monocrystalline silicon, is that of the diamond lattice. It is arranged in the form of two interwoven face-centered cubic lattices with one silicon atom placed on each lattice point [35] (Table 2, Fig. 6).

The two elementary cells are displaced against each other by a quarter of the body diagonal. The bonds are purely covalent.

Monocrystalline silicon is an indirect semiconductor. The resulting absorption coefficient is seen in Fig. 7.

Table 2 Some material properties of crystalline silicon [35, 37]

Crystal properties	
Atoms/cm^{-3}	5.0×10^{22}
Atomic weight	28.09
Density (g/cm^3)	2.24
Melting point (°C)	1,415
Vapor pressure (Pa)	1 at 1,650 °C, 10^{-6} at 900 °C
Thermal properties	
Coefficient of thermal expansion (°C^{-1})	2.6×10^{-6}
Specific heat (J/g × °C)	0.7
Thermal conductivity (W/cm × °C)	1.5
Thermal diffusion coefficient (cm^2/s)	0.9
Dielectric properties	
Dielectric constant	11.9
Electric properties	
Bandgap (eV)	1.12
Intrinsic carrier concentration (cm^{-3})	1×10^{10}
Intrinsic Debye length (μm)	24
Intrinsic specific resistivity (Ω cm)	2.3×10^5
Effective state density in the conductance band, N_c (cm^{-3})	2.8×10^{19}
Effective state density in the valence band, N_v (cm^{-3})	1.04×10^{19}
Electron affinity, χ (V)	4.05
Breakdown voltage (V/cm)	$\approx 3 \times 10^5$
Minority carrier lifetime (s) (intrinsic material)	2.5×10^{-3}
Drift mobility (cm^2/V × s) electrons/holes	1500/450

Fig. 6 Lattice structure of the silicon and the diamond lattice

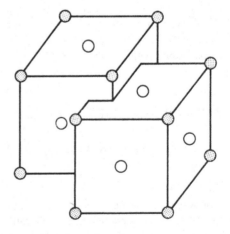

Table 1 shows a compilation of some silicon data subdivided according to crystalline, thermal, electric, and dielectric properties. As to the temperature-dependent characteristics such as thermal conductivity and dielectric properties, room temperature (300 K) is assumed.

Fig. 7 Room temperature
absorption versus wavelength
[36], slightly modified

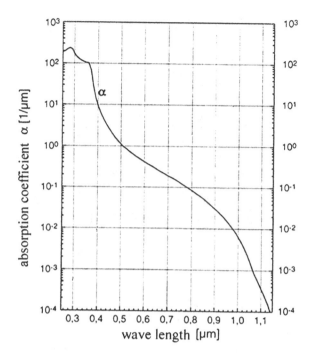

Some electric properties such as electron affinity, breakdown voltage, and drift velocity depend on doping or impurity levels so that typical values can be given only.

2.2 Useful Data of Multicrystalline Silicon

Multicrystalline (polycrystalline) silicon consists of monocrystalline grains, which are randomly oriented. The grains have sizes of some micrometer up to some centimeter and are confined by grain boundaries, regions of unfavorable electrical properties. The schematic structure and the external appearance of a multicrystalline wafer are depicted in Fig. 8.

The surface state density of the grain boundaries is 10^{11}–10^{13} V^{-1} cm^{-2}. The ratio of grain boundary diameter to wafer thickness should be larger than 5 for effective solar cell application (Table 3).

2.3 Useful Data of Microcrystalline Silicon

Microcrystalline silicon, μc-Si:H, is a material, which consists of crystallites with diameters between 1 and 100 nm. They are separated either by grain boundaries or by amorphous silicon. The electrical and optical properties of this material differ

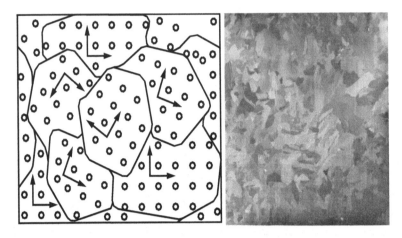

Fig. 8 Schematic representation of grains in multicrystalline material (*left*) and in multicrystalline wafer (*right*)

Table 3 Some material properties of multicrystalline silicon [38]. Note that some of these data depend on specific manufacturing and measurement conditions

Electric properties	
Specific resistivity (Ω cm)	4×10^5 (doping level $<10^{15}$ cm^{-3})
Minority carrier lifetime (s)	8×10^{-11}, grain size 0.2 µm, 7×10^{-6}, grain size 4,000 µm
Drift mobility (cm^2/V $\times$ s), electrons	40
Optical properties	
Effective refractive index	3.93 after deposition at $T < 580$ °C, 3.51 after deposition at $T > 600$ °C

strongly from those of the monocrystalline, polycrystalline, or amorphous phases [39].

Microcrystalline silicon is produced with the same setups as the amorphous hydrogen-rich material. During a-Si:H production, a small portion of the deposited film is always produced in microcrystalline form. As a rule, this contribution is negligible. However, for appropriate process parameters, the overwhelming majority of the film will appear as µc-Si:H. A typical example is the formation of the film in a plasma-enhanced chemical vapor deposition (PECVD) setup, where the silane is highly diluted in hydrogen. The latter acts as an etching gas for the amorphous phase so that the microcrystalline component is left over. Another means is the use of enhanced plasma frequencies (110 MHz instead of 13.56 MHz) [39].

Typical characteristic data and tools are resistivity, mobility, Raman, spectral ellipsometry, diffraction, and transmission electron microscopy (Table 4).

Table 4 Some material properties of microcrystalline silicon [39–41]. Note that some of these data depend on specific manufacturing and measurement conditions

Optical properties	
Index of refraction	4.05 at 500 nm
Optical band gap (eV)	1.6 at 210 °C substrate temperature
	2.0 at 350 °C substrate temperature
Absorption coefficient (cm^{-1}) at 3.5 eV	5.82 × 10^6

2.4 Useful Data of Amorphous Silicon with Respect to Heterojunction Solar Cells

Crystalline silicon (c-Si) can be designated as 4-fold coordinated network, where one silicon atom is tetrahedrally bonded to four neighboring silicon atoms. This tetrahedral structure is continued, forming a well-ordered lattice (crystal). In contrary, within amorphous silicon (a-Si), not all the atoms are 4-fold coordinated due to a continuous random network. Due to the disordered nature of the amorphous silicon, some atoms feature an unsaturated band ("dangling bond") (db). These dangling bonds can be described as broken covalent bonds, which are also found on the surface of crystalline silicon due to the absence of lattice atoms above them. The amorphous character yields either a neutral, positive, or negative charge of the db. The deviation of the bond angles averages ±10° [42]. As a result, once a metal is deposited on a silicon surface, these dangling bonds rise to interface states within the energy bandgap of silicon.

If desired, the material can be passivated by hydrogen. The bonding structure of hydrogenated amorphous silicon is depicted in Fig. 9. Additional hydrogen bonds to the db and can reduce the db density by several orders of magnitude. Thereby, SiH$_n$ groups are generated (mostly SiH and SiH$_2$), forming a-Si:H. High-quality hydrogenated amorphous silicon (a-Si:H) exhibits a hydrogen content of around 7–13 % (cf. [42, 43]), leading to a sufficiently low amount of defects within devices. Hereby, the density of states in the bandgap is reduced from 1×10^{19} cm^{-3} for a-Si (cf. [42]) down to 1×10^{16} cm^{-3} for a-Si:H(i). Thus, the "bandtails" become precipitous and the bandgap is expanded, after [44], from 1.55 eV for a-Si to around 1.7 eV for a-Si:H(i). In Fig. 10, the corresponding model of state densities after [45] is shown, denoted as Mott-CFO model (after [44]). The exponential slope of the density of states is denoted as "bandtails." The valence bandtail, with a characteristic decay-constant of 50 meV (unlike the conduction bandtail with 30 meV), determines the sub-bandgap absorption with the Urbach energy, cf. [44]. Egopt is determined by extrapolation of the delocalized states. EVm and ECm are the mobility edges. The mobility gap contains localized states in between the bandgap, while locomotive states outside the bandgap determine mainly the carrier transport. Open bonds lead to deep defect states in the middle of the bandgap. It has to be mentioned that this model appears to be a simplified model and not consistent with the present-day theory. Typical

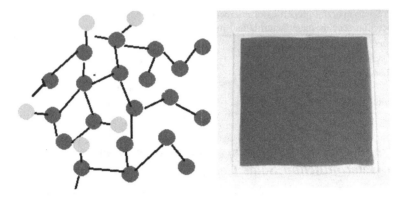

Fig. 9 Schematic representation of Si and SiH$_n$ atoms in amorphous silicon (*left*) and in amorphous layer deposited on glass (*right*)

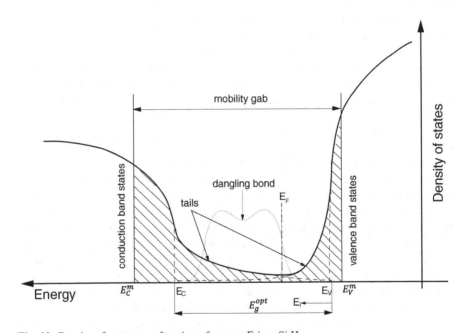

Fig. 10 Density of states as a function of energy E in a-Si:H

absorption spectrum of intrinsic a-Si:H obtained by the constant photocurrent method (CPM) and the dual-wavelength constant photocurrent method (DWCPM) is shown in Fig. 10. Additional material properties and corresponding electronic properties are given in Table 5 and can be found elsewhere [42, 46, 47].

The main predominance of a-Si compared to c-Si relies in its production technique: a very thin a-Si:H film can be deposited onto large areas by, that is, plasma-enhanced chemical vapor deposition (PECVD), while doping is achieved during plasma deposition by decomposition of doping gases to form either p- or n-

Table 5 Material properties of hydrogenated amorphous silicon [42, 48]

Optical properties	
Bandgap (eV)	1.4–1.9
Material properties	
Hydrogen content (%), typical	10–30
Density (g/cm^3) for two hydrogen concentrations	2,20 at 10 % hydrogen content
	2,11 at 20 % hydrogen content
Dielectric properties	
Dielectric constant for two hydrogen concentrations, N_H	14 at $N_H = 5 \times 10^{21}$ cm^{-3}
	6 at $N_H = 17 \times 10^{21}$ cm^{-3}
Electric properties (typical)	
Intrinsic specific resistivity (Ω cm)	2×10^{10} at room temperature
Maximum specific conductivity after doping (Ω cm)	10^{-2} at room temperature
Effective state density at the conduction band edge, N_c (cm^{-3})	4.5×10^{21}
Effective state density at the valence band edge, N_v (cm^{-3})	6.4×10^{21}
State density at midgap (cm^{-3})	10^{17}–10^{18}
Recombination lifetime (s)	10^{-9}–10^{-2}
Drift mobility (cm^2/Vs) electrons/holes	1/0.01

type layers. The resulting amorphous or microcrystalline (μc-Si:H) network, and hence the optical and electrical properties of the so deposited layers, is adjusted by PECVD process parameters. A hydrogen precursor for the a-Si:H can be silane (SiH$_4$) or additional hydrogen (H$_2$) (Fig. 11).

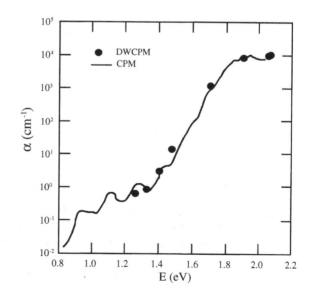

Fig. 11 Typical absorption spectrum of intrinsic a-Si:H obtained by the constant photocurrent method (CPM) and the dual-wavelength constant photocurrent method (DWCPM), [49]

3 Manufacturing

3.1 Lapping and Polishing

Bulk silicon is produced in highly pure monocrystalline or polycrystalline form. This material is sawed with a diamond saw yielding thin wafers. After that process, the silicon wafer contains saw marks and other defects on both sides of the wafer.

Lapping removes saw marks and surface defects from the front and backside of the wafers, it thins them and relieves a lot of the stress accumulated in the wafers by the sawing process [50, 51]. Another important part of the wafer manufacturing process is the edge grinding or rounding [52]. This step is done as a rule before or after the lapping. If this step is omitted, the wafer will be more susceptible to breaking in the remaining steps of the wafer manufacturing process.

After the lapping and grinding processes, the wafer is polished. Mostly, the wafer goes through 2–3 polishing steps using progressively finer and finer slurry [50, 51]. Normally, the wafer is only polished on one side, but some manufacturing processes need two-side polishing. The polishing of the surface can be omitted if the wafer is used, for example, for a textured solar cell.

3.2 Texturing

One of the main loss factors in the solar cell is the light reflection from the surface. The reflection loss of a polished silicon surface is about 34 % [36]. In order to produce high-efficiency solar cells, this reflection must be minimized (to about 10 % and lower). For the crystalline silicon solar cells, there exist two solutions: (1) an antireflection coating of the surface layer [53–56] and (2) texturing of the surface layer in combination with an antireflection coating [57, 58].

In principle, a reduction in the surface reflection by a texturing means that more light is led into the solar cell and thus more charge carriers are generated by the additional energy. However, light with a wavelength in the infrared range, for example, the thermal radiation should not be introduced into the solar cell, because this radiation only increases the temperature of the solar cell and thus reduces its efficiency. In addition to the reduction in the surface reflection, a texturing has two major goals: (1) increasing the path length of the incident light, so that more charge carrier can be generated and (2) optical trapping of weakly absorbed light by multiple internal reflections at the front and back surfaces and thereby also increasing the path length of the incident light (Fig. 12) [57–59].

The extension of the path of the incident light is particularly important for thin solar cells as these solar cells cannot fully absorb the incident light.

There are various methods of texturing a surface. In the mechanical texturing processes [60, 61], trenches are milled into the substrate. Here, the slope of the milled edge and the depth of the milled structures can be determined very

Fig. 12 Scheme for trapping light into a solar cell

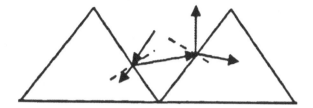

precisely. As the mechanical texturing process is independent of the crystal direction of the silicon substrate, textured substrates can be made also on poly-crystalline silicon. With this method, it is also possible to exclude certain areas of the substrate by texturing such as the busbar areas (Fig. 13). Another method is the dry chemical etching. In this method, surfaces are textured using plasma. A distinction is made between the physical plasma etching with highly accelerated ions and the chemical plasma etching with reactive gases. The usual choice in industry is the reactive ion etching (RIE) process [62–66]. This method is a combination of the two methods mentioned above, the physical and chemical etching. A significant advantage of this method consists in the fact that polycrystalline substrates can also be textured.

The most commonly used method for texturing is the wet chemical etching [58, 67, 68]. With the wet chemical etching process, the surfaces of the substrates are structured by means of acids and bases. The etching rate of these solutions is highly dependent on the crystallographic direction of the surfaces. The mono-crystalline silicon wafers which are used mainly for production of solar cells are oriented in the $\langle 100 \rangle$ crystallographic direction. In this direction, the wafers are etched several orders of magnitude faster than the $\langle 100 \rangle$ plane by means of the commonly used bases KOH (potassium hydroxide) and NaOH (sodium hydroxide). Due to the selectivity of the etching solution, pyramids are formed on the surfaces.

Fig. 13 Mechanically V-grooved surface of a silicon wafer [60]

⊢— 200 µm —⊣

Fig. 14 KOH-etched
inversed pyramids [70]

Fig. 15 Texturing with
randomly distributed
pyramids

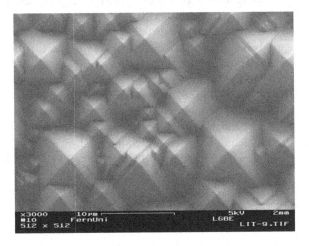

The arrangement of the pyramids on the surface can be controlled within certain limits. Inverted pyramids (Fig. 14) are formed on the surface, for example, by applying a protective grid of photoresist or SiO_2 hindering the attack of the etching solution. A texturing with inverted pyramids is used in the PERL (passivated emitter and rear locally diffused) solar cell [69]. With this concept, a cell efficiency of 25 % is achieved.

The texturing with inverted pyramids needs, as mentioned above, an additional masking step. This is associated to an increased production effort and therefore also to higher costs. Therefore, the PV industry prefers the texturing with randomly distributed pyramids (Fig. 15).

To obtain those on the surface of the substrate, the substrate is immersed in a weak (10–30 %) solution of KOH or NaOH at a temperature of 70–90 °C. The reflectance of the resulting textured surfaces is about 10 %. Newer approaches with solutions consist of even weaker concentrated (~ 2 %) KOH or NaOH solutions. In addition to these, low-concentrated solutions of isopropyl alcohol or ethylene glycol are added to improve the formation of pyramids. In other approaches, KOH or NaOH is replaced by TMAH (trimethyl ammonium hydroxide) which gives comparable results [58, 71].

3.3 Cleaning

In order to create an excellent heterointerface, it is essential to use a well-cleaned c-Si wafer surface and to remove the native oxide prior to loading the PECVD system (PECVD = plasma-enhanced chemical vapor deposition). In general, various processing techniques for cleaning can be applied prior to plasma deposition. Usually, an RCA cleaning (named after "Radio Corporation of America") is applied and followed by a dip in hydrofluoric acid (HF) in order to remove any native oxide from the wafer surface prior PECV deposition. Other works have shown (i.e., [72, 73]) that an RCA cleaning process can be avoided, if commercial wafers are used for processing, as those are already well pre-cleaned and exhibit only a native oxide of a few Angstroms. The RCA cleaning method is based on two solutions: standard-clean-1 (SC-1), an ammonium hydroxide/hydrogen peroxide/de-ionized (DI) water mixture, and standard-clean-2 (SC-2), a hydrochloric acid/hydrogen peroxide/DI water mixture. Detailed process parameters may be found in [74].

Organic compounds are removed by sulfuric peroxide; ammonium hydroxide removes particles; metallic impurities are removed by hydrochloric hydroxide; diluted HF removes the native oxide films; and for rinsing, de-ionized (DI) water is used. The mechanisms of hydrogenation of Si surfaces in fluoride-containing solutions (diluted HF) lead also to an H-termination of the c-Si surface.

After rinsing the wafer with DI water, the sample is immediately transferred into the plasma deposition chamber for deposition of a-Si:H or its counterparts by means of RF-PECVD. The time period between the last HF dip and the transfer into the PECVD chamber averages less than 5 min. It is assumed that during this procedure, the H-termination is still present and re-growth of native oxides is reduced to a minimum, cf. [72].

3.4 PECVD of i-, n-, and p-Layers

Depositions of amorphous silicon layers, whether they are intended to be used as passivation, emitter or BSF layer, or for material characterization, are usually performed in a parallel plate, mostly capacitively coupled plasma-enhanced chemical vapor deposition (PECVD) system. In this technique, a plasma occurring during the decomposition of the gaseous precursors, which can be i.e., SiH_4, H_2, CH_4, and doping gases such as PH_3 and TMB. During the decomposition inelastic collisions between high-energetic electrons and the gaseous precursor atoms result in a dissociation into atomic and ionic species. The pathways for the chemical reactions of SiH_4 and its plasma products occurring during the operation of PECVD systems can be found in [75]. For sake of simplicity, a plasma deposition as used at the LGBE, Hagen, is described in the following.

In general, the samples are placed into a 10×10 cm^2 squared sample holder, which is suitable for up to 4-inch wafer substrates, and transferred via a load lock into one of the three chambers (each chamber for either intrinsic, p-doped, or n-doped layers to prevent contamination). The sample holder is attached to the upper (electrically grounded) electrode, while the RF-power is capacitively coupled to the lower electrode. A plasma power as high as 100 W can be adjusted. The RF-PECVD system operates either at a fixed frequency of 13.56 MHz or is driven by a very-high-frequency (VHF) generator, at frequencies ranging up to very high frequencies at 110 MHz. Changing the excitation frequency does not necessarily lead to higher deposition rates, but possible changes in the micromorph structure are likely. The distance between the parallel electrodes affects the emerging network of the deposited amorphous layers and has to be adjusted precisely.

Prior to deposition, the samples are heated up before being purged with Argon gas. The Ar pressure is typically 5 mTorr for 5 min. After igniting the plasma in the chamber, a transient effect may uncontrollably influence the sample morphology in the beginning of the deposition process. Therefore, the sample holder is transferred out of the plasma field before igniting the plasma. The plasma ignition is supported by a piezo-element, after flushing the chamber with the precursors for 2 min. The deposition pressure and the gaseous precursors can be varied via the mass flow controllers (MFCs) depending on the experimental needs. The samples are radiatively heated from above with actual substrate temperatures being significantly lower than the heater temperature. Depending on the experimental conditions suitable for a-Si:H/c-Si heterojunction solar cells, the process parameters can be adjusted. The variation of those parameters includes parameters as follows:

- Plasma excitation frequency: even at the standard deposition frequency of 13.56 MHz, it is possible to achieve high-quality a-Si:H layers. However, very-high-frequency PECVD with signals of 70 or 110 MHz turns out to be more efficient and has a favorable impact on (1) the increase in the crystallinity, and (2) the growth quality of a-Si:H material.

- Deposition pressure: an increase in the deposition pressure enables a decrease in the ion bombardment, which is more preferable for the growth conditions in respect of plasma damage. However, if the deposition pressure is too high, the bias voltage increases, supporting the growth of defects in the deposited material. Moreover, high-pressure can support the formation of polysilane powder. Both effects are undesirable. Typical values of the deposition pressure are in the range of 200–500 mTorr for the PECVD setup used at LGBE, Hagen.

- Gas concentration: the gas concentration during deposition defines the optical and electrical properties, depending on the targeted film properties.

- Heater temperature (deposition temperature): those values influence the resulting network of the films drastically, as reported by Fujiwara and Kondo [76], epitaxial growth of Si occurs in the PECVD setup once a critical deposition temperature of 140 °C is exceeded. In respect of heterojunction solar cells, several authors confirmed that the interface a-Si:H/c-Si should be abrupt. Hence, epitaxial growth directly onto the c-Si wafer material would not serve as an abrupt interface.
- Plasma excitation power: an increase in the plasma excitation power influences the deposition rate, as well as a shift of the transition from amorphous to microcrystalline is observed. However, with increasing plasma power, ion bombardment increases resulting in defective material. Typical values for the plasma excitation power described in this work are in the range of 10–150 mW/cm^2.
- Electrode interspacing: the value of the electrode interspacing directly influences the either amorphous or microcrystalline nature of the deposited layers.

3.5 TCO

The transparent conductive oxide (TCO) fulfills two purposes: (1) it serves as an antireflection coating (ARC) and (2) it increases the lateral conductivity. Reflection is at a minimum when the layer thickness is an odd multiple of $\lambda_0/(4 \cdot n_{ARC})$, with λ_0 defining the free-space wavelength [77]. The ARC is therefore usually designed to present the minimum at around 600 nm, where the flux of photons is a maximum in the solar spectrum. For the preparation of heterojunction solar cells, TCO films are deposited on top of the emitter usually by DC magnetron sputtering from a ceramic target. In particular, an indium tin oxide (ITO) or a zinc oxide (ZnO) serves as TCO. The transparent conductive oxide films are fabricated at the LGBE, Hagen, using a single-chamber DC magnetron sputtering setup. The substrate temperature and the plasma power are adjusted to 200 °C and 200 W, respectively. The heater is attached in a defined distance of a few cm above the samples which are placed into a horizontal rotating table. A second horizontal rotating table gives various possibilities to operate the plasma, such as a specially designed shadow mask, a fully opened or fully closed shield. The target and the samples are separated during heat-up time by the fully closed metal shield. The sputtering time is set to 1,100 s, so that a thickness of 80 nm is obtained (corresponds to a reflection minimum at 600 nm). In Fig. 16, the typical absorption, reflection, and transmission spectra of ITO films used in this work are displayed.

3.6 Metallization and Screen Printing

The usual metallization method for solar cells in an industrial production is screen printing and subsequent firing at high temperatures of pastes consisting to a high

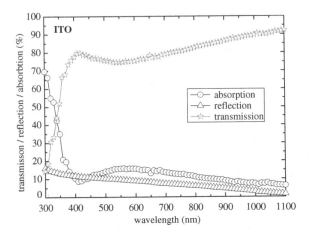

Fig. 16 Absorption, reflection, and transmission spectra of ITO deposited with 200 W at 200 °C

percentage of Ag- or Al powders. Some samples of the cross section of screen-printed Ag contact fingers are shown in Fig. 17. Additional components are organic agents to give the paste the desired viscosity, oxides which act as corrosion agent at elevated temperature and sometimes compounds containing specific doping elements. In case of p-type c-Si of course the aluminum itself acts as the doping agent creating a BSF-inducing p$^+$ c-Si layer under the Al contact. Today, the minimum width of screen-printed fingers lies below 50 μm, still much wider than structures prepared by a masking process.

High-temperature firing of these printed contact structures is getting more problematic as silicon wafers used in solar cell production are becoming thinner. The built-in tension after cooling down to room temperature results in wafer bowing, which can be so strong for thin wafers, that the solar cell can easily break during the lamination to modules. With regard to this problem, the bifacial a-Si:H/

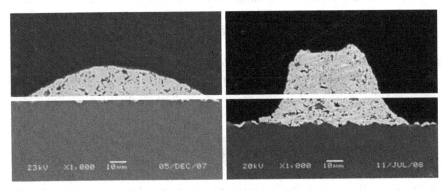

Fig. 17 Cross section of screen-printed Ag contact fingers using a conventional paste (*left*) and an optimized paste (*right*) yielding a better aspect ratio [78]

c-Si heterojunction solar cell (see Sect. 4.2) offers an alternative, because the metallization of the highly conductive TCO layers can be done with less thermal stress, since the sintering of the metal particles to decrease the track resistance of the printed metal fingers under a certain level happens at much lower temperature. Sputtering of Al contacts commonly applied in the laboratory is a low-temperature process as well and can be easily scaled up for industrial production, but wastes a lot of material, and needs much more time to reach the same aspect ratios (relation height to width of a contact finger) than screen printing.

Another low-temperature alternative is the electro or electroless plating of metals from solutions containing dissolved metal salts. Often these methods provide self-selective metal deposition only to conductive areas of the device so that no additional masking is necessary. In addition, better aspect ratios than with screen printing are possible simply by increasing the plating time [79].

4 Concepts

4.1 Conventional n a-Si:H/p c-Si Cell

Compared to a monocrystalline pn-junction solar cell, for the most basic form of an a-Si:H/c-Si heterojunction solar cell, the n-type diffused emitter is replaced by an n-type a-Si:H layer covered with a TCO layer to provide sufficient lateral conductivity, whereas the rear contact with BSF-passivation stays the same [80–82]. With this first model (Fig. 18) of a-Si:H/c-Si heterojunction solar cells, an efficiency of over 15 % has been demonstrated [83, 84].

This new emitter at the front gains two main advantages: Higher transparency of the light-entrance window, higher open-circuit voltage, due to the larger bandgap of the a-Si:H.

The finding that the recombination velocity at the a-Si:H/c-Si interface can exhibit extraordinarily low values motivated the development of p^+/p and n^+/n a-Si:H/c-Si junctions as the ohmic rear contact, avoiding at all any high-temperature diffusion process in the production scheme of a crystalline silicon solar cell.

Fig. 18 Cross section of a solar cell with TCO/n-type a-Si:H emitter at the front side of a p-type c-Si substrate

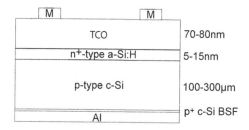

Fig. 19 Bifacial a-Si:H/c-Si
heterojunction solar cell with
intrinsic buffer layer (HIT
structure)

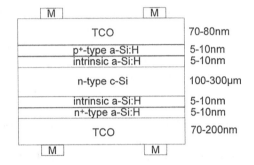

Fig. 19 Bifacial a-Si:H/c-Si
heterojunction solar cell with
intrinsic buffer layer (HIT
structure)

4.2 Bifacial a-Si:H/c-Si Heterojunction Solar Cell with Intrinsic Thin Layer, HIT Structure

Sanyo Electric Co. Ltd. introduced this double-sided a-Si:H/c-Si heterojunction solar cell in the early 1990s [85, 86]. As Fig. 19 illustrates, the superior passivation properties of the a-Si:H/c-Si interface are applied also to the rear contact, by forming an n^+ a-Si:H/n c-Si junction. The latter exhibits a back surface field effect more efficient than by a diffused p^+- or n^+-layer, since minority carriers in the c-Si absorber are reflected away from rear face recombination centers by a potential barrier in the corresponding band.

4.3 a-Si:H/c-Si Heterocontact Cell Without i-Layer

For the reduction in the interfacial recombination state density to a minimum, it appeared necessary to introduce a thin intrinsic a-Si:H layer of some nm between the c-Si base and the heavily doped a-Si:H layer. Systematic interface state monitoring during deposition series with varying doping level led to a similarly well-passivated a-Si:H/c-Si heterojunction solar cell with only one moderately doped emitter layer [87, 88]. As shown in Fig. 20, the same strategy was also applied to the rear contact avoiding the intrinsic layer between the a-Si:H layer and the c-Si base.

Fig. 20 a-Si:H/c-Si
heterocontact solar cells
similar to the structure in 4.2
but without intrinsic buffer
layer

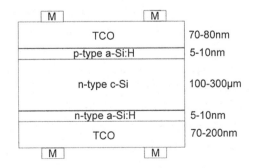

With moderately doped n-type and p-type a-Si:H layers, a good interface passivation is possible gaining an even better transparency than with additional buffer layers.

4.4 Other Concepts for Improved Entrance Windows

Another effort to increase the transparency of the entrance window is the bandgap widening of the upper layer. One possibility is the alloying of a-Si:H with carbon [89, 90].

4.5 a-Si:H/c-Si Heterocontact Cell with Inverted Geometry

The TCO/a-Si:H emitter even with a very thin intrinsic a-Si:H/p or n a-Si:H layer represents a dead layer insofar as excess charge carriers generated in these layers are lost for carrier collection due to minimum carrier lifetimes. To prevent optical losses by front side emitters, a redistribution of functionalities is necessary. Arranging the a-Si:H/c-Si interface at the rear and depositing, a very transparent a-Si_xN_y:H layer on the front side is one strategy which fulfills this task (Fig. 21) [91]. The a-Si_xN_y:H layer acts as a passivation layer to reduce the interface recombination velocity: Since the fixed positive charges in a-Si_xN_y:H induce an accumulation layer in n-type Si, the choice of n-type substrates results in a front surface field. By depositing an a-Si_xN_y:H of around 70 nm thickness, antireflection properties are added as a third functionality to this layer. The creation of the front surface field accumulation layer on the n-type c-Si substrates is the main difference compared to the front side inversion layer on p-type substrates of MIS inversion-layer silicon solar cell by Hezel et al. [92].

The a-Si:H emitter layer at the rear is therefore freed of the condition to be as thin (as transparent) as possible and can be optimized in view of minimum interfacial recombination velocity and maximized open-circuit voltage. In addition, more favorable (not necessarily transparent) conductive coatings than TCO

Fig. 21 Scheme of an p a-Si:H/n c-Si heterojunction solar cell with emitter at the rear face and highly transparent Si_xN_y window layer

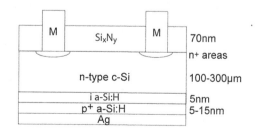

can be deposited upon the a-Si:H emitter to improve the band bending situation at the p a-Si:H/n c-Si interface. It has been shown that the TCO layer on very thin a-Si:H emitters is detrimental to the open-circuit voltage of the device, due to damages of the a-Si:H/c-Si interface during a non-sufficiently soft TCO deposition [93] or because of the potential drop between the TCO and the doped a-Si:H emitter reaching through the emitter and thereby lowering the open-circuit voltage at the a-Si:H emitter/c-Si interface [94].

The ohmic contact at the front face has to be prepared through the a-Si$_x$N$_y$:H layer. Among the many options, laser ablation of a-Si$_x$N$_y$:H, selective electroless plating of the openings, and local laser-annealing appear to be the solution with the lowest front grid shading and the best maintenance for the electronic passivation of the a-Si$_x$N$_y$:H/c-Si interface [79].

4.6 Interdigitated HIT Cell

The final goal for any photovoltaic device is to avoid at all shading or filtering of the sunlight entering the cells absorber. This of course necessitates the preparation of both contacts at the rear and the use of an extremely transparent passivation layer at the front as already used for the inverted cell described above. Up to now, the only commercially available back contact solar cell is produced by SunPower using diffused n$^+$ and p$^+$ contact regions. The self-aligned isolation process between the n$^+$ and p$^+$ regions important for the good fill factor of these cells has to be replaced by much more sophisticated procedures if n-type and p-type a-Si:H layers have to be prepared side by side. Several approaches to such IBC (Inter-digitated Back Contact) solar cells with heterojunctions have been undertaken by many research group working on a-Si:H/c-Si heterojunction solar cells [95–97]. The schematic of the structure is shown in Fig. 22.

Fig. 22 Cross section of one possible interdigitated back contact HIT structure proposed [98, 99]

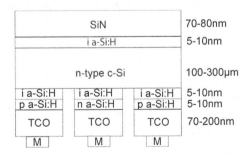

5 Problems and Challenges

5.1 Choice of the Base Material, Impact of the Doping, n/p Versus p/n

There are a large number of parameters which characterize the base material: Growth, orientation, doping, surface treatment, area, thickness, doping density, and type in a first group, defects, lifetime, and surface recombination velocity in a second group, just to name the most important of them. The parameters of the first group are easily controllable while the parameters in the second group are difficult to improve.

Most laboratories use Czochralski (CZ) grown materials. The reason is that this material contains fewer defects than float-zone (FZ) or multicrystalline material and therefore higher efficiencies are obtained [100].

An important parameter of heterojunctions is the relative distribution of the hetero-band offset between valence and conduction band (cf. Fig. 23). While most authors hypothesized a major contribution of the band offset in the valence band [10, 101], also the contrary has been proposed, based on XPS measurements [102]. This controversy has an analogy in the history of the GaAlAs/GaAs heterojunction, where in earlier times, the Dingle rule [103] assumed a 85:15 % distribution of the hetero-band offset between conduction and valence band, while later on a 60:40 % distribution has been found [104]. This shows that even in classical III–V semiconductor heterojunctions, the interfacial properties are not easy to determine. An excellent discussion of this history is given by Kroemer [105]. As a consequence, it was recognized quite early that the p-type a-Si:H emitter on n-type substrates led to distinctly higher open-circuit voltages and hence higher efficiencies compared to n-type a-Si:H emitters on p-type substrates [106]. This is due to the appearance of the larger band offset in the valence band than in the conduction band. The latter results in a more favorable band bending situation on both sides of a a-Si:H/c-Si heterojunction solar cell with n-type absorber: In the case of p-type a-Si:H on n-type c-Si (right), the built-in potential $\Phi_b{}^p$ is larger than $\Phi_p{}^b$ in the inverse case, n-type a-Si:H on p-type c-Si resulting in a larger usable open-circuit voltage for this configuration.

Fig. 23 Energy band diagram of the two anisotropic a-Si:H/c-Si heterojunctions [107]

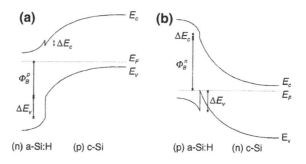

On the other hand at the back contact, the offset in the conduction band is smaller for an p^+-a-Si:H layer on a p-type substrate than the valence band offset for an n^+ a-Si:H layer on an n-type substrate representing the higher back surface field for holes in n-type c-Si than for electrons in p-type c-Si.

5.2 Surface States

Surface states (also termed interface states) are well known in MOS (metal-oxide-semiconductor) electronics. The interface in question is that between the oxide and the semiconductor. In the 1960s and 1970s, they were a severe obstacle to the operation of MOSFETs, CMOS inverters, and related circuitry. In order to reduce them by technological means, one had to be able to determine their density first. A lot of measurement techniques—for example, bridge measurements—were developed at this time, and technologies for their reduction—for example, application of atomic hydrogen—were investigated. Interestingly enough, the results of both have successfully been applied to surface states on the solar cell substrate. Now, the investigated surface is that between the wafer and the a-Si layer.

A typical bridge measurement on a heterojunction solar cell exhibiting a low surface state density is shown in Fig. 24.

This dispersion is easily explained by means of the underlying equivalent circuit shown in Fig. 25.

In this figure, C_1 and R_1 stand for the junction capacitance and its loss, and R_3 stands for the bulk resistance. The low-frequency series resistance is the sum of R_1 and R_3, while the high-frequency series resistance is just given by R_3. A similar discussion holds for the series capacitance.

However, when the substrate is replaced by a lower-grade material, the frequency dispersions of Fig. 24 distinctly change. As an example, we show the capacitance dispersion in Fig. 26.

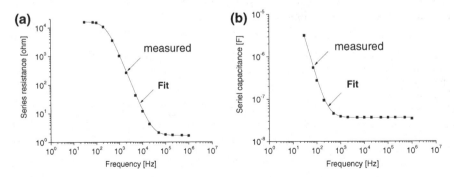

Fig. 24 Resistance and capacitance of an a-Si:H/c-Si solar cell as measured in the series mode

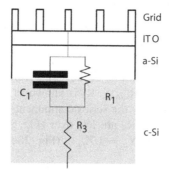

Fig. 25 Heterojunction solar cell structure and one-time constant equivalent network

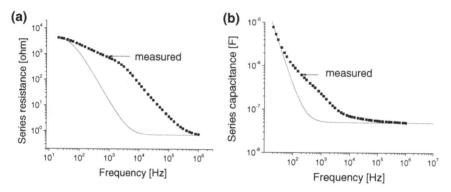

Fig. 26 Resistance and capacitance of an a-Si:H/mc-Si solar cell as measured in the series mode. The full lines reflect the equivalent network of Fig. 25

An additional signal in the range of 50 Hz–1 MHz is superimposed to the "ideal" frequency dependence of a single time constant model. The explanation is the incorporation of states at the a-Si:H/m-Si interface. Let us suppose that a continuum of surface states has been built in the a-Si:H/m-Si interface (Fig. 27).

It is evident that the equivalent circuit of Fig. 25 must be amended by the contribution of the surface states. They offer an additional recombination path in

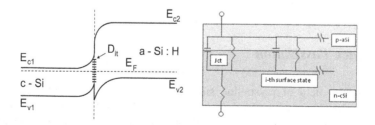

Fig. 27 Cross section of the heterojunction with surface states (*left*) and equivalent circuit (*right*)

parallel to the junction. Each energy level E_i in the forbidden band introduces one C–R time constant.

In the flow diagram of Fig. 28, the way to check the model is described [108]. The measured network reflects the contribution of the junction, C_1 and $G_1 = 1/R_1$, of the interface states, C_{it} and G_{it}, and of the series resistance, R_3 (Fig. 28a). In a first step, R_3 is subtracted from $R_{s,meas}$. R_3 is known as the high-frequency series resistance of Fig. 28a. Thus, only the series components $Z_{s,corr}$ between the upper terminal and the T-point above R_3 are left over (Fig. 28b). Now, these data are converted to a parallel network, $Y_p = 1/Z_{s,corr}$ (Fig. 28c). The Y_p components are seen in Fig. 29a and b.

At this stage, we make use of the interface admittance properties from Eqs. (1) and (2) [109]

$$\frac{G_{it}}{\omega} = \frac{qD_{it}}{2\omega\tau_{it}} \ln(1 + \omega^2\tau_{it}^2) \tag{1}$$

$$C_{it} = \frac{qD_{it}}{\omega\tau_{it}} \arctan(\omega\tau_{it}) \tag{2}$$

It is deduced that the surface state conductance disappears for $f \to 0$ and the surface state capacitance disappears for $f \to \infty$ so that the measured conductive component (Fig. 29a) and the capacitive component (Fig. 29b) are identical to $G_1 = 1/R_1$ and C_1. Finally, the admittance $Y_{it} = Y_p - G_1 - j\omega C_1$ is obtained (Fig. 28d).

The resulting components G_{it}/ω and C_{it} are plotted in Fig. 30a, b, respectively. A fit is applied by means of Eqs. (1) and (2) so that a D_{it} of 2.3×10^{12} V^{-1} cm^{-2} and a time constant τ_{it} of 1.7 ms are obtained. The results are the same for the conductance and the capacitance evaluation.

In conclusion, it is found that multicrystalline silicon is much richer in interface states than monocrystalline material. It can be surmised that the reason for this difference is the inherent nature of the multicrystalline grain structure. The true locations of the states are the grain boundaries in the vicinity of the interface. The defects at these locations might be either intrinsic structural defects or foreign impurities attracted by the grain boundaries. A good means to reduce these effects to a minimum is seen in an early surface passivation step.

Several other groups have investigated capacitance and conductance in silicon heterojunctions using various modes [80, 110–112].

5.3 Surface Passivation

In the research field of crystalline silicon (c-Si) solar cells, electronic surface passivation has been recognized as a crucial step to achieve high conversion efficiencies. High bulk and surface recombination rates are known to limit the open-circuit voltage and to reduce the fill factor of photovoltaic devices [112,

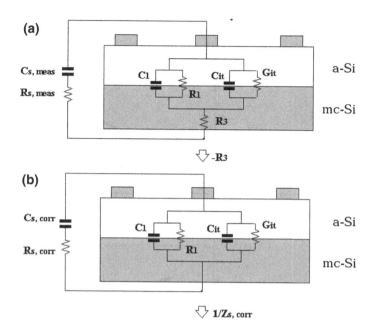

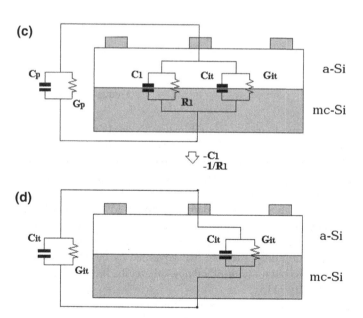

Fig. 28 Flow diagram of the interface admittance evaluation. The network at the *left* (outside the sample) shows the mode (parallel or serial) on which the data are based. In (**a**), the total network is shown. In (**b**), the series resistor R_3 is subtracted and $R_{s,corr}$, $C_{s,corr}$ are *left*. In (**c**), the network is transformed to a parallel representation. Finally, the space charge components are subtracted so that only the surface state contributions are seen (**d**)

Fig. 29 **a** Parallel
capacitance after correction
for the bulk resistance.
b Parallel conductance after
correction for the bulk
resistance

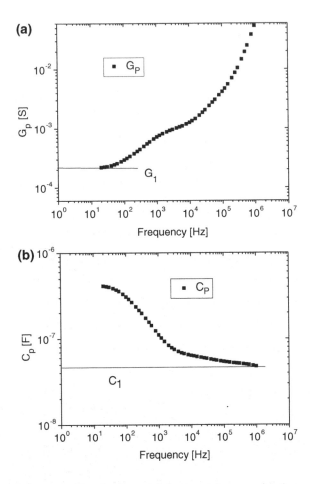

113]. The suppression of surface recombination by applying a surface passivation scheme is thereby one of the basic prerequisites to obtain high-efficiency solar cells. This becomes particularly true for heterojunction solar cells: featuring an abrupt discontinuity of the crystal network at the crystalline silicon (c-Si) surface to the amorphous emitter (a-Si:H) usually results in a large density of defects in the bandgap due to a high density of dangling bonds [113]. These defects at the heterointerface often induce detrimental effects on the solar cell performance, cf. [114]. Although the electrical field can reduce the recombination near the heterointerface, the junction properties are still governed by the interface state density. Therefore, in order to obtain high-efficiency solar cells, it is essential to reduce the interface state density [115].

Passivation schemes commonly used in photovoltaic applications utilize silicon dioxide (SiO_2) [116], and silicon nitride (SiN_x) [117, 118], but also intrinsic amorphous silicon (a-Si:H(i)) [119], and amorphous silicon carbide (a-SiC:H) [120].

Fig. 30 a Normalized interface state conductance (G_{it}/ω) versus frequency. The dotted line shows the fit according to a continuum interface state model. **b** Interface state capacitance (C_{it}) versus frequency. The dotted line shows the fit according to a continuum interface state model

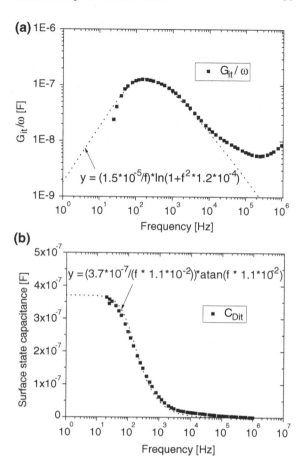

5.3.1 SiO₂ and SiNₓ Passivation Schemes

As described earlier, standard passivation schemes include high-quality SiO_2 and SiN_x films for high-efficiency device fabrication. Up to now, record values of effective lifetimes are established for SiO_2 and SiN_x by Kerr and Cuevas published in [116, 117]: the highest effective lifetimes (τ_{eff}) which incorporates all the bulk and surface recombination processes, previously reported for crystalline silicon appears to be $\tau_{eff} = 29$ ms and $\tau_{eff} = 32$ ms, corresponding to the lowest surface recombination velocity (SRV) S_{eff} with 0.46 and 0.625 cm/s for a 90 Ω cm FZ n-type material and a 150 Ω cm FZ p-type material, respectively, passivated with annealed SiO_2. Note that the values given are obtained using the effective lifetime at its maximum value, neglecting the choice of a discretionary injection level. However, the SRV for higher doped material, that is, 1.5 and 0.6 Ω cm n-type material, increased to 2.4 and 26.8 cm/s, respectively, for wafers passivated with SiO_2. It is well known that the values for the measured τ_{eff} and the corresponding S_{eff} strongly depend on the chosen value for the mean carrier density (MCD), and the c-Si wafer doping and resistivity used for the investigation of passivation quality.

Thermally grown silicon oxide has shown excellent surface passivation properties, resulting in a very low state density. However, the growth implies a high-temperature application ($\sim 1{,}050$ °C), and it suffers from long-term UV instability. Low-temperature processing sequences are based mainly on passivation with silicon nitride (SiN_x) [117], amorphous silicon films [119], or amorphous silicon carbide [120] or stacks of those. The SiN_x films are silicon rich, and this fact brings along several drawbacks: the passivation quality depends strongly on the Si doping type and level used; the films show a considerable absorption in the ultraviolet range of the solar spectrum, leading to a reduction in the J_{sc}; the etch rate of those films is extremely low, hindering the local opening of the SiN_x, which makes them non-applicable for heterojunction solar cells.

5.3.2 a-SiC:H Passivation Scheme

Passivation schemes using a-SiC:H are intensively investigated by, that is, Vetter et al. [90] and Martin et al. [120]. An a-SiC:H passivation of 3.3 Ω cm p- and 1.5 Ω cm n-type c-Si results in $\tau_{eff} > 1$ ms. Using the same model as Garin et al. [121], it leads to the identification of a field-effect passivation mechanism: fixed positive charge for p-type c-Si, fixed negative charge for n-type c-Si. Vetter et al. [122] investigated the effect of amorphous silicon carbide layer thickness on the passivation quality of crystalline silicon surfaces.

5.3.3 a-Si:H Passivation Scheme

Excellent surface passivation abilities of non-doped a-Si:H(i) films on c-Si have been intensively investigated by other research institutes and confirmed by various methods, cf. [123–125]. Dauwe et al. [126] published very low surface recombination velocities on p- and n-type silicon wafers passivated with hydrogenated amorphous silicon films. Silicon material of both conduction types can be effectively passivated. Wang et al. [127] investigated a-Si:H/c-Si heterointerfaces, stating that they necessitate an immediate a-Si:H deposition and an abrupt and flat interface to the c-Si substrate.

Especially for heterojunction solar cell applications, a-Si:H(i) has attracted the photovoltaic community due to the success of HIT (heterojunction with intrinsic thin layer) cells [25, 115, 128]. The a-Si:H(i) films can be grown by PECV deposition at low temperatures (<200 °C). However, Fujiwara and Kondo [129] stated that the growth of a-Si:H at temperatures $T_{dep} > 130$ °C often leads to an epitaxial layer formation on the c-Si, reducing the solar cell performance. Also, due to the inherent strong blue light absorption, only ultra-thin a-Si(i):H films can be allowed to prevent losses. Physical and structural properties of a-Si:H, however,

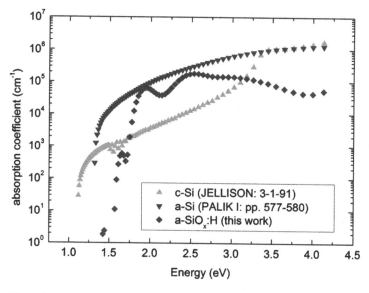

Fig. 31 Absorption coefficient of a-SiO$_x$:H deduced from SE data compared to published absorption coefficients of crystalline silicon and a-Si:H. The absorption coefficient at the blue light region around 3.0 eV is by one order of magnitude lower than the one of standard a-Si:H. Taken from [130, 131]

strongly vary with the growth conditions, and the characteristics of the a-Si:H/c-Si heterointerface still remain under discussion.

5.3.4 a-SiO$_x$:H Passivation Scheme

Focusing on the junction fabrication techniques of a-Si:H/c-Si solar cells using a low-temperature PECVD technique, passivation of the surface regions of the cell avoiding high-temperature cycling becomes an important issue. High-transparent PECV-deposited hydrogenated amorphous silicon suboxides (a-SiO$_x$:H), grown at low temperatures, represent a material system suitable for this application and are quite an attractive alternative to standard a-Si:H incorporated in a-Si:H/c-Si heterojunction solar cells [130, 131]. Record high effective lifetime values of 4.7 ms on 1 Ω cm n-type FZ wafers and 14.2 ms on 130 Ω cm p-type FZ wafers prove the surface passivation applicability of silicon wafers to any kind of silicon-based solar cells. The values achieved in this work appear to be the highest ever reported values of τ_{eff} on a high-doped crystalline wafer of 1 Ω cm resistivity. Additionally, the a-SiO$_x$:H films yield a surface passivation quality exceeding earlier published record passivation schemes such as SiN$_x$ and SiO$_2$. Therefore, the use of a-SiO$_x$:H may be a promising alternative for any passivation scheme existing so far. Advantages of the a-SiO$_x$:H passivation scheme are that the fabricated a-SiO$_x$:H layers are grown by simple PECV deposition at low temperatures, they withstand hydrofluoric acid (tested with a 5 % diluted HF dip for 1 min) and high temperatures up to 350 °C. It has to be mentioned that the process parameters have to be carefully adjusted to obtain such high-quality a-SiO$_x$:H films. A high-transparent window layer up to 2.4 eV, depending on the oxygen fraction in the precursor gas, is obtained by adding oxygen to the a-SiO$_x$:H layers. The a-SiO$_x$:H films exhibit a very low blue light absorption compared to a-Si:H(i) (cf. Fig. 31), enabling a thicker passivation layer than standard a-Si:H(i) and therefore an increasing passivation quality.

Furthermore, the applicability of a-SiO$_x$:H layers as a high-quality passivation scheme used in heterojunction solar cell fabrication is an effective alternative to standard a-Si:H(i) buffer layers. It is found that the surface passivation quality depends on the deposited layer thicknesses. Particular importance of that parameter becomes evident for the application of those films in the production of heterojunction solar cells.

The key factors to obtain high-quality a-SiO$_x$:H films comprehend (1) the chamber contamination, which depends on previous processes with various precursor gases, (2) dummy process to cover the reactor walls with a-SiO$_x$:H, (3) the deposition temperature and the preheating time in the chamber, and (4) the electrode distance.

5.4 PECV-Deposited Emitter and Back Surface Field

5.4.1 Hydrogenated Amorphous Silicon (a-Si:H)

The requirements of high optical depth and perfect charge collection imply very high demands of material quality. Thus, in heterojunction solar cell devices, a-Si:H has been accepted as a suitable heterojunction material forming both emitter and surface passivation layer. Nevertheless, the light absorption in the a-Si:H(i) and the doped a-Si:H layers is rather strong, and the short-circuit current density in the a-Si:H/c-Si solar cells decreases rapidly with increasing a-Si:H layer thicknesses [132]. To improve the short-circuit current density in a-Si:H/c-Si solar cells further, it is preferable to employ an a-Si:H-based alloy that has a larger optical bandgap than standard a-Si:H. It is well known that for, that is, thin-film solar cells, the PECVD conditions required to produce high-quality a-Si:H are close to the region where microcrystalline Si is deposited. With increasing crystallinity, the doping efficiency and carrier mobility increase. However, for the growth of those layers directly on c-Si, often an epitaxial growth at the interface is observed, reducing the effective lifetime and the V_{oc} of those devices [133].

5.4.2 Wide-Gap, Hydrogenated Amorphous Silicon Carbide Films (a-SiC:H)

Hydrogenated amorphous silicon (a-Si:H) has been accepted as a suitable heterojunction material for a-Si:H/c-Si solar cells, and conversion efficiencies exceeding 22 % have been reported recently [134]. Nevertheless, photon absorption in the emitter of a heterojunction solar cell leads to a considerable current loss with increasing a-Si:H layer thickness due to the high recombination in this layer. To improve J_{sc} in a-Si:H/c-Si solar cells further, it is preferable to employ an a-Si:H-based alloy that has larger optical bandgap than a-Si:H to suppress light absorption in the window layer or to reduce the recombination.

Amorphous alloys of silicon and carbon (a-Si$_x$C$_{1-x}$:H$_y$) are a promising alternative to standard a-Si:H. The introduction of carbon adds extra freedom to control the properties of the resulting material. An increasing concentration of carbon in the alloy (x) is used to widen the electronic gap between conduction and valence bands, in order to potentially increase the light efficiency of solar cells made with amorphous silicon carbide layers [135]. However, besides the ability to enhance the optical bandgap and suppress absorption in the thin-film emitter, it changes the electronic properties of hydrogenated amorphous silicon: the electronic semiconductor properties (mainly electron mobility) are badly affected by the increasing content of carbon in the alloy, due to the increased disorder in the atomic network. Bringing (x) to the opposite extreme of carbon concentration (100 %) results in amorphous carbon, or synthetic diamond-like films. Several studies are found in the scientific literature, mainly investigating the effects of deposition parameters on electronic quality (cf. [136–139]), but practical applications of amorphous silicon carbide in commercial devices are still lacking, in particular the use of a-Si$_x$C$_{1-x}$:H$_y$ in heterojunction solar cell devices.

The influence of deposition parameters on the incorporation efficiency of both hydrogen and carbon in a-Si$_x$C$_{1-x}$:H$_y$ and a-Si$_x$:H$_y$ films for use as emitters in heterojunction solar cells has been studied by Mueller et al. [140]. It is found that the optical bandgap E_G of those films can be tailored from 1.5 eV up to 2.3 eV with an appropriate addition of both methane as a carbon source and hydrogen during PECV deposition. The μ-Raman spectra analyses have shown that additional methane as carbon source in the precursor material broadens the band at $\sim 2{,}000$ to $\sim 2{,}100$ cm^{-1}, which can be attributed to the formation of SiH$_2$ species. Compared to additional hydrogen without methane in the feedstock, only SiH-related bonds are visible, indicating the typical a-Si:H spectra at $\sim 2{,}000$ cm^{-1}. Addition of both, hydrogen and methane, shows an enhanced intensity and enlarged broad band from 2,000 to 2,150 cm^{-1}, indicating trihydride SiH$_3$ bonds and the incorporation of C in the form of CH$_3$ groups. Thus, Raman spectra at ~ 840 and ~ 890 cm^{-1} indicate carbon incorporation, which can be attributed to a-SiC. However, no C–C bonds are visible, which would lead to low conversion efficiency due to the sp^2/sp^3 bonding structure. Spectra in the high wavenumber range at $\sim 2{,}900$ cm^{-1} exhibit peaks related to C–H bonds due to the high methane content in the precursor material. Two sample series are investigated in more detail: (1) 15 % methane with a defined hydrogen concentration in the precursor material and (2) 30 % hydrogen with a defined methane concentration. For these sample series, the stretching modes of Si–H and C–H bonds are analyzed by decomposing the spectra into sub-peaks around $\sim 2,000$ and $\sim 2,100$ cm^{-1}. Heterojunction solar cells, prepared with an appropriate addition of hydrogen and carbon, show an increased V_{oc} as well as an increased J_{sc}. From IQE measurements (in the wavelength range from 350 to 550 nm), one can conclude that the addition of hydrogen leads to a slightly better internal quantum efficiency. The addition of carbon, however, decreases the IQE slightly in the short-wavelength region. As a consequence, the conversion of photon energy into electric current implicates that it is not performed efficiently by just adding carbon, due to the sp^2/sp^3 bonding structure of C–C [113, 138, 139] as well as to the decreasing photoelectronic properties, such as photoconductivity. Despite of that the dark conductivity of the resulting thin films decreases with enhanced addition of carbon to the precursor material. The trade-off between electrical defect density and optical absorption of the emitter layer results in a bandgap of approximately 2.0 eV and a layer thickness of 5–10 nm. Light in the long-wavelength region passes this layer almost without loss; the optical losses in the blue spectral range are typically around 10 %. The V_{oc} results of the cells discussed in this section appear to be quite poor; however, this is an expected behavior for any p–n or n–p heterojunction structure, as the interface state density caused by the doping materials, which attach to the c-Si surface during the deposition process, appears to deteriorate the junction properties significantly, cf. [115]. By the addition of carbon, the optical bandgap widens so that the suppression of light absorption in the window layer is reduced, but the density of defect states increases. Hydrogen dilution minimizes this deterioration. From our I–V measurements and the quantum efficiency measurements, it is evident that the addition of carbon and hydrogen does contribute to the photogenerated current. It is recommended that by proper valency control of the optical bandgap due to the hydrogenated amorphous carbon–silicon alloys, a-Si$_x$C$_{1-x}$:H$_y$ J_{sc} and V_{oc} should be developed to a maximum efficiency of the solar cells. Also, with a non-doped a-Si:H layer sandwiched between the heterojunction, the heterointerface is separated from the doped layer so that these defects caused by the doping materials might be avoided. This implies that the essence of the heterojunction solar cells consists in creating a good interface to avoid surface recombination [115], which has been discussed in Sect. 5.3.

5.4.3 Wide-Gap, Hydrogenated MicroCrystalline Silicon Films (μc-Si:H)

According to the film properties discussed in this section, doped (p$^+$ and n$^+$) μc-Si:H films are likely to be suitable for use as emitter and BSF in a heterojunction solar cell device. They indicated high transparency to suppress absorption, and high conductivity when annealed at the optimum temperature. Results on the influence of those films compared to standard a-Si:H by Mueller et al. can be found in [141].

It is found that the fraction of microcrystallinity can be varied as a function of hydrogen dilution. Low hydrogen dilution results in a larger fraction of amorphous structures inside the deposited film, whereas high H_2 dilution (>98 %) results in a more microcrystalline character. The μc-Si:H(p) layers indicated high conductivities of $\sigma_{dark} = 10$ S/cm (corresponding to 0.1 Ω cm) at optimal PECV deposition parameters. The n$^+$ μc-Si:H films indicated significant higher dark conductivity values (with optimum process parameters approximately $\sigma_{dark} = 100$ S/cm) compared to the σ_{dark} values of p$^+$ μc-Si:H films. It can be concluded that the fraction of microcrystallinity can be varied with the plasma frequency, and is also influenced by the deposition temperature and hydrogen dilution. The doping efficiency and hence the carrier mobility are optimal at VHF plasma frequency of 110 MHz, and with high hydrogen dilution of $\chi H = 98$ %.

The dark conductivity σ_{dark} and carrier mobility are further improved by means of thermal annealing for both n$^+$ and p$^+$ μc-Si:H films. The μc-Si:H(p) films annealed with temperatures in the range of 200 °C $< T_{ann} < 375$ °C are appropriate to obtain the high dark conductivity values (with optimum process parameters results in approximately $\sigma_{dark} = 25$ S/cm). By means of post-annealing of the μc-Si:H films or a-Si:H films with a microcrystalline fraction, an activation of the boron atoms occurs, which was formerly bound in form of B–H-Si. The annealing step might release the hydrogen from this B–H-Si bound, leaving activated B–Si bounds. For μc-Si:H(n) films, an annealing temperature in the range between 180 °C $< T_{ann} < 300$ °C is found to be an optimum temperature to improve dark conductivity up to $\sigma_{dark} = 130$ S/cm (corresponding to 0.0077 Ω cm).

6 Measurement Techniques

6.1 Absorption, Reflection, and Transmission

In general, the incident solar radiation (I) must equal reflected (R) plus transmitted (T) plus absorbed (A) radiation.

$$I = R + T + A \tag{3}$$

For the a-Si:H layers as well as for the transparent conductive oxides (TCO), maximum transmission is intended. Consistently, the fraction of light transmitted to the wafer which is available for solar conversion increases. Since the TCO act as an antireflection coating, the knowledge of the reflection is also mandatory. Thus, the usual techniques are employed. Most common techniques are mainly based using the Ulbricht sphere (more details of the set up are given in Sect. 6.5.2), spectroscopic ellipsometry, or a spectrophotometer. An example of those measurements for a specific TCO, namely ITO, is given in Fig. 16, displaying the absorption, reflection, and transmission spectra of ITO deposited on glass.

6.2 Excess Charge Carrier Lifetime

The foremost problem in Si-based solar cells is the suppression of recombination processes. A very sensitive tool for the characterization of the charge carrier kinetics in silicon in general but especially for the evaluation of electronic passivation of Si-interfaces is the contactless conductivity measurement, such as the interaction with megahertz magnetic fields (QSSPC), with microwaves (μ-WPCD, TRMC, TRPC, FRMC...), terahertz pulses or sub-bandgap infrared light. For an introductory example, let us consider microwave radiation (1–100 GHz) being reflected from a semiconducting sample which is terminating a waveguide system, which is a good choice with regard to sensitivity and shielding of parasitic external influences (Fig. 32).

Depending on the kind of excitation, the detector is selected accordingly and its signal is either fed to a fast transient recorder [142], a lock-in amplifier [143] or a sensitive voltmeter [144]. Electromagnetic radiation in the megahertz or gigahertz frequency range interacts strongly with semiconducting electrodes via free charge carriers. Therefore, changing the number of free carriers within the sample by injection processes (electrically via contacts, thermally, optically, or radiantly with high-energy particles (charged or neutral)) should result in different absorption and reflection properties for electromagnetic radiation, as depicted in Fig. 32.

Microwave power is generated in a Gunn oscillator and is directed via an isolator, an attenuator, and a circulator toward the sample, which terminates the waveguide system. The reflected microwave power is directed by the circulator to the detector diode. For time-resolved measurements, laser pulses induce excess charge carriers in the sample and the increased conductivity is followed by the microwave power reflected from the sample. If light with a small absorption coefficient is used a homogeneous excess charge carrier profile in a silicon sample symmetrically passivated with a 70 nm SiN$_x$-layer results in an exponential decay of the conductivity as illustrated in Fig. 33.

Some theoretical considerations given in the following are helpful for the interpretation of experimental results.

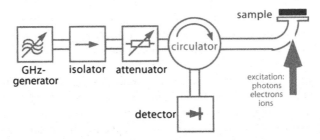

Fig. 32 Microwave reflectometer composed of waveguides, active and passive devices for a specific frequency band

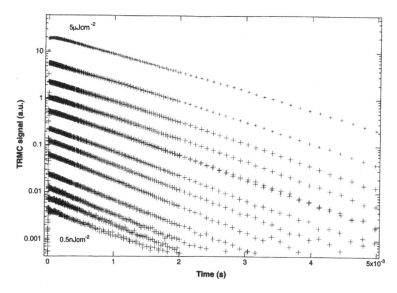

Fig. 33 Typical decay of photoexcited (1,064 nm, 10 ns FWHM) excess charge carriers in a bifacially nitride-passivated n-type c-Si (140 Ω cm, 525 μm) sample measured with a microwave reflectometer at 10 GHz as shown in Fig. 32. [145]

For small perturbations, it can be shown that the change in reflected microwave power ΔP_r is proportional to the stationary reflected power P_r [16, 146] and the changes ΔC_e, ΔC_h of the numbers of excess electrons and holes in the sample:

$$\Delta P_r(t) = S P_r(\mu_e \Delta C_e(t) + \mu_h \Delta C_h(t)) \tag{4}$$

The contributions of electrons and holes are weighted with their mobilities μ_e and μ_h. If no traps for electrons or holes are present and a net charging of the sample due to charge transfer of excess carriers into contacting phases under open-circuit conditions can be excluded, $\Delta C_e(t) = \Delta C_h(t)$ holds, since $\Delta C_e(t)$ and $\Delta C_h(t)$ are spatially integrated excess concentrations of electrons and holes. The proportionality factor or "sensitivity factor" S accounts for the lateral distribution of the microwave field over the sample, the dark conductivity of the sample, and several geometric effects. If the sensitivity factor is varying spatially (i.e., the electric component of the microwave field is varying over the samples volume), the simple redistribution of carriers in the sample, that is, the change in the shape of a given charge carrier profile can induce a shift in the ratio of reflected and absorbed microwave power. Therefore, care has to be applied to ensure well-defined experimental conditions.

The value of the sensitivity factor, which is assumed to be time independent for small perturbations, can be quantified empirically under certain experimental conditions [147], but in most cases, the evaluation of the relative change in the microwave signal with experimental parameters is sufficient.

In the following, gains and losses of excess minorities (holes for the present consideration) should be discussed with regard to excess charge carrier measurements. The number (excess concentration integrated over the sampled volume V) of generated minority carriers per time ΔC_h^g is balanced by the numbers of carriers lost per time via transfer ΔC_h^{tr}, via bulk recombination ΔC_h^{br}, and via surface recombination ΔC_h^{sr} (Fig. 34):

$$\frac{d\Delta C_h(t)}{dt} = \Delta C_h^g(t) - \Delta C_h^{br}(t) - \Delta C_h^{sr}(t) - (\Delta)C_h^{tr}(t) \tag{5}$$

with $\Delta C_h(t) = \int_v \Delta c_h^{ss}(x,y,z,t)dxdydz$.

Since the dimension of the above terms is s^{-1}, Eq. (5) describes an equilibrium of (particle–) currents.

In a fast transient measurement, for sufficiently short excitation pulses, the loss processes can be assumed to be inactive during excitation:

$$\Delta C_h^{br}(\tau_{pulse}) + \Delta C_h^{sr}(\tau_{pulse}) + \Delta C_h^{tr}(\tau_{pulse}) = 0. \tag{6}$$

Then, the maximum ΔP_r^{max} of the transient ΔP_r-signal is proportional to the integral of the generation over the duration τ_{pulse} of the exciting light pulse, assuming the independence of the minority carrier mobility on time:

$$\Delta P_r^{max} = S^* \mu_h \int_0^{\tau_{pulse}} \Delta C_h^g(t)dt \tag{7}$$

The contribution of the majority carriers is taken into account by a modified sensitivity factor S^*, which includes P_r as well. With Eq. (7), the determination of the charge carrier mobility is possible.

Whereas in a stationary measurement, Eq. (6) becomes

$$\Delta C_h^g = \Delta C_h^{br} + \Delta C_h^{sr} + \Delta C_h^{tr} \tag{8}$$

with

$$\Delta C_h^{br} = s_b A \overline{\Delta c_h^{ss}} = s_b A \frac{1}{V} \iiint_v \Delta C_h^{ss}(x,y,z)dxdydz = \frac{s_b}{d} \Delta C_h^{ss}$$

$$\Delta C_h^{sr} = s_r A \overline{\Delta c_h^{ss}(x,y,z=0)}$$
$$= s_r A \frac{1}{A} \iint_A \Delta C_h^{ss}(x,y,z=0)dxdy = s_r \iint_A \Delta C_h^{ss}(x,y,z=0)dxdy$$

$$\Delta C_h^{kr} = k_r A \overline{\Delta c_h^{ss}(x,y,z=0)}$$
$$= k_r A \frac{1}{A} \iint_A \Delta C_h^{ss}(x,y,z=0)dxdy = k_r \iint_A \Delta C_h^{ss}(x,y,z=0)dxdy$$

For a sample with volume v and with the steady-state excess concentration $\overline{\Delta C_h^{ss}}(x, y, z = 0)$ at the surface (averaged over the illuminated area $\mathcal{A}$ of size A) and the normalized average excess concentration $\overline{\Delta C_h}^{ss}$ in the bulk.

In this case, ΔP_r is proportional to the steady-state number of excess charge carriers ΔC_h^{ss} which is not only determined by generation ΔC_h^g, but also by the effective recombination velocities in the bulk (s_b), at the surface (s_r) and the effective charge transfer velocity (k_r):

$$\Delta P_r = S^* \mu_h \Delta C_h^{ss} = S^* \mu_h \iiint_v \Delta C_h^{ss}(x, y, z) \mathrm{d}x \mathrm{d}y \mathrm{d}z \qquad (9)$$

For laterally homogeneous excitation and recombination over the area A, the spatial integration in (9) reduces to

$$\Delta P_r = S^* \mu_h A \int_0^d \Delta C_h^{ss}(z) \mathrm{d}z \qquad (10)$$

For a given excitation ΔC_h^g, a decrease in ΔC_h^{ss} can be caused by an increase in the quantum yield or an increase in the recombination velocities. Therefore, the stationary microwave signal is complementary to the sum of all three loss components.

For the theoretical description of the microwave signal ΔP_r, the remaining task is the determination of the stationary excess minority carrier profile $\Delta c_{ss}(z)$ and its dependence on the recombination and transfer conditions at the boundaries of the sample. The latter has been worked out for semiconductor electrolyte interfaces in [147, 148].

Under the conditions given above in time-resolved measurements, the decay of the measured microwave power ΔP_r reflected from the sample is caused by the decaying excess charge carriers (integral in Eq. 10). The decay can be characterized with an effective decay time. In an ideal case, each of the loss currents of Eq. 5 is correlated with a single decay process characterized by a decay time τ.

$$\frac{1}{\tau_{\mathrm{eff}}} = \frac{1}{\tau_{\mathrm{bulk}}} + \frac{1}{\tau_{\mathrm{surface}}} + \frac{1}{\tau_{\mathrm{emitter}}} \qquad (11)$$

Depending on the kind of semiconductor and the excitation level, τ_{bulk} is the result of different recombination processes such as radiative recombination, band-to-band recombination, Auger recombination, and Shockley–Read–Hall recombination via midgap states. In silicon under low injection conditions, only the latter mechanism has to be considered. τ_{surface} accounts for recombination via surface states, and τ_{emitter} represents all losses associated with an emitter layer on top of the bulk absorber. These can be the recombination of excess carriers in that layer but also the collection (given by k_r in Fig. 34 and Eq. 8) of photogenerated carriers if the charge transfer from the absorber to the emitter layer comes along with a drop in mobility [149].

Fig. 34 Charge carrier balance in an excited semiconductor sample

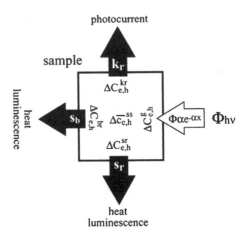

Changing the boundary conditions (i.e., surface recombination or charge transfer velocity) changes the ratios of ΔC_h^{sr}, ΔC_h^{tr}, ΔC_h^{br} and ΔC_h^{ss}. If the system can be driven in a regime where one of these decay processes dominates, measuring the product of this decay process (i.e., photocurrent, luminescence, and heat), beside the microwave reflection, yields the respective kinetic constant under these conditions, provided there is no other process resulting in the same decay product.

According to Sinton [150], interpreting the effective lifetime depends on the kind of sample: the impact of all recombination paths depends on the excess carrier density, Δn, because the distinctive recombination path in a given sample may vary depending on the injection level (Δn). That means different recombination parameters might be extracted at different injection levels. Samples which are of interest in this work are (1) silicon wafers either passivated symmetrically by an isolating layer, for example, silicon suboxide silicon nitride or intrinsic amorphous silicon and (2) silicon wafers with plasma-deposited doped regions (emitter) forming junctions such as pn diodes or back surface field structures (p^+p, n^+n).

Two special but very common sample structures should be specified in order to simplify the applicable as further.

- *sample structure (i)*

For a sample with no charge collecting interfaces, the measured effective lifetime is then a combination of bulk and surface recombination

$$U_{\text{eff}} = U_{\text{bulk}} + U_{\text{surface,front}} + U_{\text{surface,back}} \tag{12}$$

$$\frac{\Delta n}{\tau_{\text{eff}}} = \frac{\Delta n}{\tau_{\text{bulk}}} + \frac{S_{r,\text{front}}\Delta n_s}{W} + \frac{S_{r,\text{back}}\Delta n_s}{W} \tag{13}$$

For a symmetrically prepared ($s_{r,\text{front}} = s_{r,\text{back}} = s_r$) silicon sample of thickness W, and if the bulk lifetime is sufficiently long to allow generated carriers to reach both surfaces, and if s_r is sufficiently low, the effective carrier lifetime is limited by the surface recombination velocity [151] as

$$\frac{1}{\tau_{\text{eff}}} = \frac{1}{\tau_{\text{bulk}}} + \frac{2S_r}{w} \qquad (14)$$

For good quality silicon wafers, bulk recombination s_b can be neglected compared to surface recombination s_r.

The other limiting case is a very high surface recombination velocity. Then, the excess charge carrier transport from the bulk of the sample to the surface is the limiting factor for the decay of photoexcited conductivity. Therefore, the characteristic surface decay time constant is expressed by a relation containing only ambipolar diffusion constant and sample thickness

$$\frac{1}{\tau_{\text{eff}}} = \frac{1}{\tau_{\text{bulk}}} + \frac{D_{n,p}\pi^2}{w^2} \qquad (15)$$

where $D_{n,p}$ is the ambipolar diffusion constant, with $D_n = 28 \text{ cm}^2/\text{s}$ for 1 Ω cm p-type c-Si and $D_p = 11 \text{ cm}^2/\text{s}$ for 1 Ω cm n-type c-Si, cf. [152].

- *sample structure (ii)*

For samples symmetrically deposited with highly doped regions, such as an emitter or a BSF structure on a silicon wafer, the measured effective lifetime can then be determined from the relation

$$\frac{\Delta n}{\tau_{\text{eff}}} = \frac{\Delta n}{\tau_{\text{bulk}}} + \frac{J_{o,\text{front}}np}{qn_i^2 w} + \frac{J_{o,\text{back}}np}{qn_i^2 w} \qquad (16)$$

where n_i is the intrinsic carrier concentration in silicon. This expression can be further simplified to

$$\frac{1}{\tau_{\text{eff}}} = \frac{1}{\tau_{\text{bulk}}} + \frac{J_{0,\text{front}}(N_{\text{dop}} + \Delta n)}{qn_i^2 W} + \frac{J_{0,\text{back}}(N_{\text{dop}} + \Delta n)}{qn_i^2 W} \qquad (17)$$

where N_{dop} defines the background dopant density and J_{front} or J_{back} the emitter or BSF saturation current density, respectively [150].

6.2.1 Quasi-Steady-State and Transient Photoconductance Decay Via Inductive Coupling

The determination of excess charge carrier decay via an inductive coupling of the sample to a high-frequency magnetic field has been frequently used due to affordable equipment by Sinton Consulting. With its WCT-120 system, two different measurement modi are possible:

A detailed description of the QSSPC and the TPCD method and the setup used in this work can be found in [153, 154]. After Nagel et al. [155], the carrier density can be expressed as

$$\frac{d\Delta n(t)}{dt} = G(t) - U(t) + \frac{1}{q}\Delta J \tag{18}$$

For the case of a homogeneous distributed carrier density, or homogeneous generation rate G, the last term in Eq. 18 is negligible. Bearing in mind that $U_{eff} = \Delta n/\tau_{eff}$, the effective lifetime can be written as

$$\tau_{eff} = \frac{\Delta n(t)}{G - \frac{d\Delta n(t)}{dt}} \tag{19}$$

This expression reduces to the transient expression in case $G = 0$, and to the QSS case when $d\Delta n/dt = 0$.

QSSPC

As for the QSSPC method, the carrier lifetime can be obtained via the photo-conductance based on the following method. The silicon wafer sample is exposed to a light pulse which is exponentially decaying with a time constant considerably longer than the excess charge carrier decay processes to be measured. If the excess carrier populations are close to be in "steady state," then the generation and recombination rates are in balance. The excess carrier density, Δn, can be achieved by measuring the sheet conductivity as a function of time inductively coupled to the sample by an RF-coil as shown in Fig. 35. Then, each moment in time corresponds to different injection levels. The measured sheet conductivity can then be used to calculate the excess carrier density (Δn) using existing models for carrier mobilities.

The intensity of the flash is measured simultaneously with a calibrated photodiode (Fig. 35) as a function of time and converted into a generation rate, G, of electron–hole pairs in the sample. This requires the knowledge of an optical constant, an estimate of the amount of incident light that is absorbed in the sample though, which is mainly based on optical models on known values of the absorption coefficients and refractive indices of a silicon wafer and various surface films. Sinton [150] gives a detailed description about possibilities obtaining the optical constant, using the software PC1D [156]. With the known generation rate and the excess charge carrier density determined from the measurement, the effective lifetime can be calculated via a steady-state condition:

$$\Delta n = G\tau_{eff}.Z \tag{20}$$

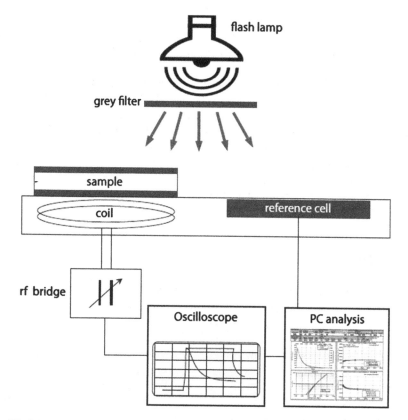

Fig. 35 Schematic of the inductively coupled photoconductance apparatus used for the effective lifetime measurements

TPCD

The transient PCD method does not require a value for the optical constant, but contrary to the QSSPC method, this technique is only appropriate for the evaluation of photogenerated carrier lifetimes appreciably longer than the flash turnoff time. In the TPCD mode, the sample is subjected to shorter light pulses than in the QSSPC-mode. A time frame of typically 10–20 μs is used for the excitation. After the flash finished, and the carriers have redistributed homogeneously to a steady-state profile across the wafer, the decay of sheet conductivity can be observed as a function of time. The effective carrier lifetime at different carrier densities is determined via $\tau_{\text{eff}} = -\Delta n/(d\Delta n/dt)$. Figure 35 illustrates a schematic of the used setup for QSSPC and TPCD measurements.

Combining the Two Measurement Modi

Evaluating the measured effective lifetime over the widest possible range of the excess carrier density (ECD), measurements of QSSPC/TPCD data analyzed in the corresponding QSS, generalized, or transient mode, are combined. After *Kerr and Cuevas*, the strength in using both QSSPC and TPCD methods to determine the effective carrier lifetime is that while the two methods are complementary, they also have different dependencies on the system calibration constants. The TPCD method is quite independent of the calibration constants for a system with linear response, while the QSSPC method relies on them. An excellent overlap in the lifetime measurements as a function of injection level for the two methods is therefore a good verification that the calibration constants were accurately determined.

6.2.2 In Situ TRMC

Beside the contactless methods described above which all give integrated information over a more or less extended sample volume, they can also be applied to map the lateral homogeneity of optoelectronic properties [157] and can be used as a powerful in situ tool for process control [19].

Thin-film semiconductor growth is a rather complex process, which is influenced by a large number of parameters. Therefore, the application of in situ measurement techniques is of great interest. There have been strong efforts to correlate the measured plasma parameters and the chemical precursor distribution in the gas phase with the deposited film properties [158]. Another possible approach for in situ process control and parameter optimization is direct measurements of the growing film properties. In this latter case, non-invasive techniques that do not require special test structures are certainly of great advantage. Optical techniques, like spectroscopic ellipsometry [159, 160], have been shown to give detailed information regarding the film structure. Microwave reflection–based techniques can be used to measure important semiconductor parameters like charge carrier mobilities and minority carrier lifetimes. And they are common tools for silicon wafer inspection before processing. Here, we will demonstrate the capabilities of transient photoconductivity measurements, based on the microwave reflection change in semiconductor films after generation of free carriers by short laser pulses [16], for the in situ characterization during PECVD growth of amorphous hydrogenated silicon (a-Si:H) and for the kinetics of the formation of various interfaces.

During the formation of heterojunctions for solar cells, based on amorphous silicon deposited on crystalline silicon substrates, transient microwave–detected photoconductivity measurements permit to follow the kinetics of the initial crystalline silicon surface damaging and the subsequent defect passivation during amorphous silicon deposition in real time. A newly developed simulator enables us to model also the fast initial decay of the microwave transients [161].

The experimental setup consists of a time-resolved microwave conductivity (TRMC) system that has been hooked up to a conventional capacitively coupled RF glow discharge PECVD system operating at 13.56 MHz. Process gases are silane, hydrogen, phosphine, and diborane, enabling the deposition of intrinsic, n-type, and p-type amorphous hydrogenated silicon (a-Si:H) at typical substrate temperatures between room temperature and 300 °C. A schematics of the complete system is given in Fig. 36. The TRMC system consists of a tunable Gunn oscillator operating between 26.5 and 40 GHz (Ka-Band), a circulator, and a fast microwave detector diode. The constant microwave power (P) of the Gunn oscillator is directed through the circulator to the open end of a waveguide, that is conducted through the lower RF-electrode, and covered with a vacuum window on top of which the substrate—for the reported measurements: crystalline silicon—is fixed. A Nd:YAG laser is used to illuminate the surface of the growing thin-film semiconductor with 10-ns-long pulses at wavelengths of 532 and 1,064 nm. Typical laser intensities are in the order of 100 μW cm^{-2} for the characterization of amorphous silicon and of 1 μW cm^{-2} in the case of crystalline silicon. The photogenerated charge carriers change the thin-film conductivity and hence the microwave reflection. The reflected microwave power (P + ΔP(t)) yields therefore information regarding the excess charge carrier kinetics. The reflected microwave is returned to the circulator and then directed toward the microwave detector diode. The resulting output signal is recorded in a fast transient recorder and subsequently stored in a personal computer.

When we grow amorphous silicon on top of a crystalline silicon substrate, up to appreciable a-Si:H film thicknesses, the main contribution to the TRMC-signal, when we generate excess charge carriers by illumination through the growing a-Si:H layer, is related to the charge carriers generated in the crystalline silicon substrate. This is due to the large asymmetry in the charge carrier mobilities between the two materials. Using a wavelength (for example 532 nm) where the light is also strongly absorbed in the crystalline silicon, the TRMC technique can

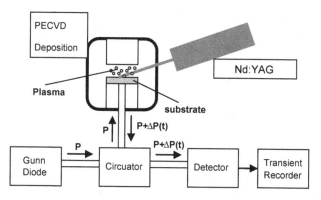

Fig. 36 Schematics of the TRMC—measurement setup for the in situ characterization of the PECVD growth of amorphous silicon layers

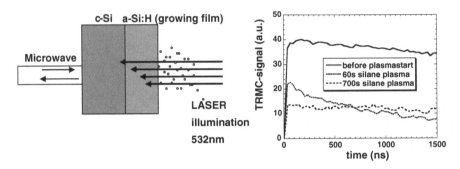

Fig. 37 **a** Measurement configuration and **b** TRMC transients measured at 250 °C before and during the growth of an intrinsic a-Si:H layer on top of a p-type crystalline silicon substrate after excitation with a 10 ns laser pulse at a wavelength of 532 nm

monitor the charge carrier kinetics right at the interface between the two materials. The experimental configuration is given in Fig. 37a.

Typical transients measured before and after different times after silane plasma ignition are shown in Fig. 37b on a short time scale. This short time scale has been chosen, in order to follow the changes in the charge carrier kinetics with a good time resolution. There is of course a trade-off between time resolution and signal-to-noise ratio of the TRMC transients.

We observe in Fig. 37b that the TRMC transient before plasma start is only slowly decreasing at 1,500 ns after pulsed laser illumination to about 80 % of the initial value. The transient measured 60 s after plasma start exhibits lower amplitude and is much faster decaying. Both are due to the plasma damage of the crystalline silicon substrate that results in a strongly enhanced surface recombination. After 700 s of silane plasma exposure, the signal amplitude is further decreasing but the decay rate is comparable to the value before plasma start. The lower amplitude can be explained by an increased number of absorbed photons in the growing a-Si:H film, while the slower decay is due to a passivation of the surface by the growing film. In Fig. 38, we see the development of the TRMC amplitude and the decay time as a function of the a-Si:H deposition time. The expression "relative lifetime" refers to the signal value at 1,500 ns divided by the signal amplitude. The TRMC amplitude decreases, besides the initial drop, exponentially with increasing a-Si:H deposition time, and we can use the rate of the amplitude decrease in the signal in order to determine the absorption coefficient at 532 nm of the a-Si:H film.

The transient measured 60 s after plasma start has clearly a lower amplitude and is much faster decaying. Both are due to the plasma damage of the crystalline silicon substrate that results in a strongly enhanced surface recombination. After 700 s of silane plasma exposure, the signal amplitude is further decreasing but the decay rate is comparable to the value before plasma start. The lower amplitude can be explained by an increased number of absorbed photons in the growing a-Si:H film, while the slower decay is due to a passivation of the surface by the growing

Fig. 38 Development of the
TRMC-signal amplitude and
relative lifetime during the
deposition of an intrinsic a-
Si:H layer grown at 250 °C
on top of a p-type crystalline
silicon substrate after
excitation at 532 nm

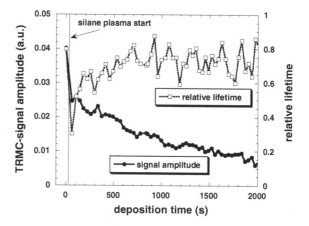

film. In Fig. 38, we see the development of the TRMC amplitude and the decay
time as a function of the a-Si:H deposition time. The expression "relative lifetime"
refers to the signal value at 1,500 ns divided by the signal amplitude. We see that
the TRMC amplitude decreases, besides the initial drop, exponentially with
increasing a-Si:H deposition time, and we can again determine the absorption
coefficient at 532 nm of the a-Si:H film.

The relative lifetime is dropping initially to a rather low value and completely
recovered after 1,000 s of a-Si:H deposition. It has been shown that the kinetics
and the degree of the passivation are, for example, strongly dependent on the
substrate temperature as one key parameter of the a-Si:H deposition process [20].

6.3 Electroluminescence

Another characterization technique that is rather sensitive to interface modification
of solar cells is the electroluminescence (EL) under forward bias. Zhao et al. [162]
found that the solar cells with the highest conversion efficiency performed also best
as infrared light emitters. This opens, for example, the doors to the fabrication of all-
silicon optocouplers. The electroluminescence technique became very popular in
the last years for the defect imaging of mono- and polycrystalline silicon solar cells
because it is non-invasive and easy to install not only in a laboratory, but also in a
production environment [163]. Applied at room temperature, the main information
is obtained by the amount of the suppression of the band-to-band emission in
crystalline silicon due to defect-related non-radiative recombination [164, 165]. It
should, however, be mentioned that even more detailed information can be obtained
by the electroluminescence due to defect-related new radiative transitions at longer
wavelengths. These—in general weaker electroluminescence signals—are much
easier detected at cryogenic temperatures [166].

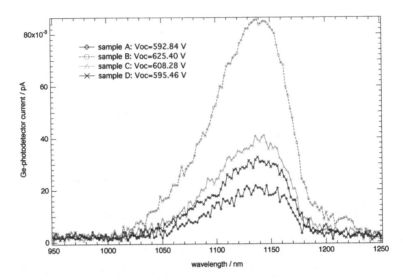

Fig. 39 Electroluminescence spectra of ITO/a-Si:H/a-SiO:H/n c-Si Heterojunction solar cells for a-SiO:H buffer layer thicknesses varying in the range from 2 nm to 4 nm for samples A–C (Structure symmetric on both sides. Sample D was prepared with a thinner a-SiO:H layer at the rear. The corresponding open-circuit voltages of these cells are given in the legend

So far, photoluminescence [167] and electroluminescence [168, 169] measurements have been applied to the a-Si:H/c-Si heterojunction in order to detect defects and to investigate the heterointerface.

For the spectrally resolved measurements, shown in Fig. 39, a setup very similar to the one shown in Fig. 42 has been used, with the only difference that the sample (solar cell) is placed as a forward-biased light-emitting diode at the position of the Xe arc lamp and the InGaAs or Ge photodiode is placed at the position of the sample in Fig. 42 and connected to the lock-in amplifier.

6.4 a-Si Characterization

6.4.1 Ellipsometry

In general, ellipsometry refers to the measurement of the change in polarization state of light reflected from the surface of a given sample. The measured values are expressed as Ψ and Δ and derived from the determination of the relative phase change in a beam of reflected polarized light. These values are related to the ratio, ρ, of Fresnel reflection coefficients R_p and R_s for p- and s-polarized light, respectively, cf. [170, 171]. For sake of simplicity, a variable angle spectral-ellipsometer (VASE) (*J.A. Woolam, INC.*) with a spectral range from 250 to 1,700 nm as used at the LGBE, Hagen, is described in the following. The setup consists of a rotating analyzer ellipsometer (RAE) in which the light beam leaves the monochromator, passes through a *fixed* (input) polarizer, is reflected from the sample, passes through a *rotating* polarizer (the analyzer, which is continuously rotating) and then strikes the Si detector. In general, the detector signal is measured as a function of time, the measured signal is Fourier analyzed in order to obtain the Fourier coefficients a_n and b_n, and finally, Ψ and Δ are calculated from a_n and b_n and the known azimuthal angle, P, of the input polarizer.

Ellipsometry measurements provide the best results for a film thickness in a wavelength range of the light used for the measurement. Also, roughness features on the sample surface or film interface should be less than 10 % of the probe beam wavelength for the ellipsometric analysis to be valid. More detailed information about ellipsometer physics and the VASE can be found in [171]. Ellipsometry as an optical technique requires an accurate model of the measurement process to analyze the input data. In many cases, the spectral acquisition range and angles of incidence allow the deduction of both thickness and optical constants of the same film. The key components of the ellipsometric models are the optical constants of the substrate and sample layers and the thickness of the layers. The optical constants are parameters which characterize how matter will respond to excitation by an electromagnetic radiation at a given frequency.

The real part or index of refraction (n) defines the phase velocity of light in material. The imaginary part or extinction coefficient (κ) determines how fast the amplitude the wave decreases. The extinction coefficient is directly related to the absorption of a material. Thus, the optical constants expressed by the real and imaginary parts of the complex index of refraction (n and κ) represent the optical properties of a material in terms of how an electromagnetic wave will propagate in that material. Alternatively, the real and imaginary parts of the dielectric function contain the same information in terms of how the material responds to an applied electric field [171].

The absorption coefficient, α, determines how far into a material light of a particular wavelength can penetrate before it is absorbed. Its value can be deduced from the extinction coefficient κ after Davis and Mott [45].

To determine the optical bandgap E_G of the amorphous or microcrystalline plasma-deposited films, the dispersion model for amorphous material based on the absorption edge *Tauc* formula by Davis and Mott [45], and the quantum mechanical Lorentz oscillator model, proposed by Jellison-Jr. and Modine [172], has been employed.

Usually, polished crystalline wafer or corning glass is used as substrate, as roughness feature on the sample surface or at the film interface should be generally less than 10 % of the probe beam wavelength for the ellipsometric analysis to be valid. In case a c-Si substrate is used, the SiO_2 layer introduced serves as an optical separation layer. An accurate fit of very thin films benefits from a sandwiched thermally grown silicon dioxide layer between crystalline wafer and amorphous/microcrystalline plasma-deposited layer to enhance the resolution.

The SE data are fitted assuming a three-layer model (unless stated differently): *surface roughness/a-Si:H film/corning glass 7,059*, or a four-layer model *surface roughness/a-Si:H film/SiO₂/c-Si.* The optical bandgap is determined from ellipsometry data using mostly the Tauc method, whereas a number of different conventions have been established among different authors; and a comparison between those pinpoints a slight difference of E_G.

6.4.2 Raman

Micro-Raman (μ-Raman) spectroscopy is used for the observation of changes in the microscopic network of for instance a semiconductor due to growth conditions or other treatment. To understand the Raman effect, a short overview of Raman theory will be given here. Detailed information can be found in the literature [173–177].

When monochromatic radiation impinges upon a molecule of a semiconductor, it may be reflected, absorbed, or scattered in some manner. Light scattered from a molecule has several components—the Rayleigh scatter, and the Stokes and anti-Stokes Raman scatter (approximately only 1×10^{-7} of the scattered light is Raman scatter). *Rayleigh scattering* designates a process without any change in frequency. *Raman scattering* specifies a change in frequency or wavelength of the incident radiation. The spontaneous Raman effect occurs, when the incident photons with the energy of $h\nu_0$ thus interact with that molecule and excite an electron from the ground state to a virtual energy state. Relaxing into a vibrational excited state generates *Stokes Raman scattering*. The Raman-shifted photons can be either of higher or lower energy, depending upon the vibrational state of that molecule. Since there are a small number of molecules, which are existent in an elevated vibrational energy level, the scattered photon can actually be scattered at a higher energy. In this case, the Raman scattering is then called *anti-Stokes Raman scattering*. It is the amount of energy change by that photon (either lost or gained), which is characteristic of the nature of each bond (vibration) present and provides the chemical and structural information.

Not all vibrations will be observable with Raman spectroscopy (depending upon the symmetry of the molecule). Hence, the amount of energy shift for a CH bond is different to that seen with a Si–H bond. By analyzing all those various wavelengths of scattered light, it is possible to detect a range of wavelengths associated with the different bonds and vibrations.

The crystalline volume fraction measured via Raman spectroscopy is a criterion to describe the silicon materials in its transition zone from amorphous to crystalline. In Raman spectroscopy, the standard peaks for crystalline and amorphous silicon appear at 520 and 480 cm^{-1} wavenumbers, respectively. Therefore, the position of the Raman peaks for microcrystalline silicon determines prevalence of either crystalline or amorphous structure in the deposited films. Under these considerations, the peak at 500 cm^{-1} indicates dominance of crystalline fraction in the examined samples, whereas the position of the peak close to 480 cm^{-1} attributes amorphous character of microcrystalline film. The terminology of microcrystalline is not accurately defined in literature, as one can use the term "microcrystalline layer" or "close to the transition to crystallinity" for a PECV-deposited layer, which contains only a fraction of microcrystals.

6.4.3 Conductivity

The measurement of the dark and photoconductivity allows one to evaluate the basic electronic properties (such as doping concentration, with the knowledge of the carrier mobility know from additional techniques such as Hall effect, or Time Of Flight (TOF)) of PECV-deposited a-Si:H layers and its alloys. The dark conductivity (σ_{dark}) and photoconductivity (σ_{ph}) of the deposited a-Si:H films and its counterparts are determined using the standard four-point probe method as well as the *transfer length method* (TLM). In the four-point-probe method, a current passes through two outer probes, whereas the generated voltage is measured only through the inner probes, which allows the measurement of the substrate resistivity. Detailed information can be found elsewhere [178]. The TLM is based on a measurement technique originally proposed by Shockley; the method consists of current–voltage characteristic measurements, using a coplanar electrode configuration as illustrated in Fig. 40 schematically. The implementation of this method implies unequally spaced contacts (as displayed). The voltage is measured between the contacts to ensure there is only bare semiconductor between the contacts; no other contacts should interfere with the measurement. The plot of the total resistance as a function of the contact spacing d can be derived once the values of R_T are measured for various contact spacing. From this plot, it is possible to extract (1) the *sheet resistance* by the slope $\Delta(R_T)/\Delta(d) = \rho_s/Z$, (2) the *contact resistance* by $R_T = 2R_c$ at the intercept at $d = 0$, and (3) the *specific contact resistivity* with ρ_s known from the slope of the plot by $-d = 2L_T$ at the intercept at $R_T = 0$.

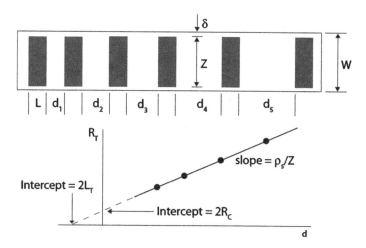

Fig. 40 Transfer length method (TLM) test structure and a plot of total resistance as a function of contact spacing, d. After [178]

6.4.4 Deep-Level Transient Spectroscopy

Deep-level transient spectroscopy (DLTS) is a technique for monitoring and characterizing deep levels introduced intentionally or occurring naturally in a-Si:H. In particular, the method can determine the activation energy of a deep level, its capture cross section and concentration, and can distinguish between traps and recombination centers.

The DLTS technique is mostly used to observe the thermal emission from majority carrier traps, which can be recorded as a capacitance or voltage transient. The basic principle of a DLTS measurement is as follows. The bias on a diode is pulsed between a bias near zero and some reverse bias V_R with a repetition time t_r as shown in Fig. 41a. The zero bias condition is held for a time t_f during which the traps are filled with majority carriers in this interval the capacitance signal contains no useful information and the instrument may be overloaded by the high zero bias capacitance. During the reverse bias pulse, the trapped carriers are emitted at a rate e_n producing an exponential transient in the capacitance, which in its general form can be written as

$$C(t) = C(\infty) + \Delta C_o \cdot \exp(-t/\tau_e) \qquad (21)$$

Fig. 41 Diagram indicating the principle of a DLTS measurement: Part (a) illustrates the repetitive filling and reverse bias pulse sequence, and part (b) shows the diode capacitance transient, as a function of time. Part (c) indicates the variation of the transient time constant τ $(=e_n^{-1})$ with reciprocal temperature for two different traps, and Part (d) illustrates the deep-level transient spectrum that is produced by a rate window with reference time constant τ_{ref} operating on the capacitance transient shown in (b)

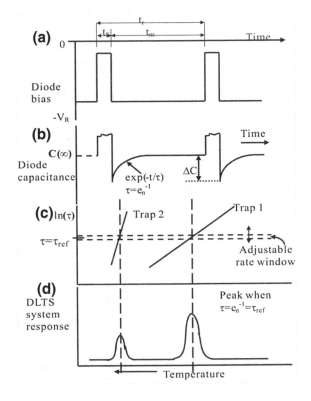

The time constant τ_e is equal to e_n^{-1}, and ΔC_o is given by Eq. 21. This transient is illustrated in Fig. 41b. The essence of the DLTS method is to feed this transient to a "rate window" that provides a maximum output when the time constant τ_e is equal to a known preset time constant τ_{ref}. The operating principle of a DLTS measurement is as follows. Consider a test diode on material containing two different traps, each characterized by a trap signature in the form of a linear plot of $\ln(\tau)$ versus T^{-1} with different values of Ea and $\sigma\infty$ as shown in Fig. 41c. As the temperature of the diode is increased, the emission rate increases, and a peak appears in the rate window output as $\tau = e_n^{-1}(T)$ passes through τ_{ref} for each trap. This is depicted in Fig. 41d.

For a given τ_{ref}, the peak temperatures T_1 and T_2 are characteristic for each trap and the peak height is proportional to ΔC_0 and gives the trap concentration N_T. Moreover, by repeating the scan with different values of τ_{ref}, sets of values of e_n and T can be obtained. In turn, the activation energy E_a and capture cross section σ_n of each trap can be determined from Arrhenius plots of $\ln(\tau_n T^2)$ versus T^{-1} [179].

6.5 Electronic Device Characterization

6.5.1 Current–Voltage Characteristic (IV)

The most fundamental of solar cell characterization techniques is the measurement of the cell efficiency. Standardized testing allows the comparison of devices manufactured at different companies and laboratories using various technologies.

The *standard test conditions* (STC) for solar cell characterization are (1) air mass 1.5 g spectrum (AM1.5 g) for terrestrial cells, (2) intensity of 1,000 W/m² (100 mW/cm², one-sun illumination), (3) cell temperature of 25 °C, and (4) four-point probe technique to remove the effect of probe/cell contact resistance. A typical setup to carry out current–voltage (IV) characteristics is described in [180], and commercially available from different companies. The STC for solar cells described above have been strictly adhered to in the framework. For example, a deviation in temperature introduces errors in the values of V_{oc}. One-sun illumination is quite intense; therefore, the cell is placed on a commercially available, water-cooled brass block. A thermocouple is inserted in the block and the control system is adjusted so that the required temperature of 25 °C is kept constant. To contact the cell using the four-point probe technique, a current and voltage probe on top of the cell and a current and voltage probe on the bottom of the cell are used. The metal block acts as the rear (current and voltage) contact. The top contacts are usually paired since it is insufficient to have a single voltage and current probe. Ideally, the probes make good contact with the cell and the voltage and current probes are within close proximity but not touching each other [181].

6.5.2 Spectral Response

While the quantum efficiency specifies the electron output of the cell compared to the incident photons, the spectral response (SR) is defined as the ratio of the current generated by the cell to the incident power on the cell surface. The I_{sc} is measured under monochromatic illumination with the intensity $\Phi_0(\lambda)$. The SR is then defined as $SR(\lambda) = Isc(\lambda)/\Phi_0(\lambda)$. Usually, the spectral response is measured in units of [A/W]. A typical SR setup is illustrated in Fig. 42. The quantum efficiency (QE) can be calculated based on the SR measurements: the quantum efficiency is determined from the spectral response by replacing the power of the light at a particular wavelength with the photon flux for that wavelength. The QE can be defined as the ratio of the number of carriers collected by the cell to the number of photons of the incident light.

A solar cell under illumination at short circuit generates a photocurrent depending on the incident light. The photocurrent density (J_{sc}) can be related to the incident spectrum via the cell's quantum efficiency (QE). Therefore, the QE is a key quantity to describe the solar cell performance under different conditions [182].

The external quantum efficiency (EQE) does not depend on the incident spectrum, so to say light that does not enter the cell, or light that leaves the cell again after entering, does not contribute to the photocurrent. Thus, the EQE depends upon the absorption coefficient of the solar cell material (expressed by the probability, f_{abs} that an incident photon, carrying the energy E, will be absorbed) and the efficiency of charge separation and charge collection in the device (enunciated by the probability f_c). The EQE of a solar cell includes the effect of optical losses such as transmission and reflection. A quantity that depends less strongly on the optical design (or a quantum efficiency of the light left after the reflected and transmitted light has been lost) can be given by the internal quantum efficiency (IQE). The IQE is taking the reflectance data, R, of the device into account to obtain a corrected EQE curve. The IQE gives hereby the probability

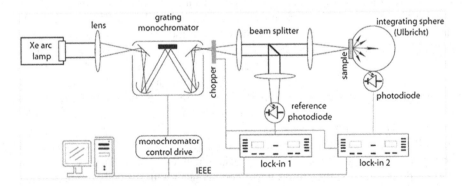

Fig. 42 Schematic illustration of setup for transmission measurements. The same setup is used for reflection and spectral response measurements; therefore, the arrangement of the sample can be re-adjusted depending on the intended measurement

that an incident photon of a certain wavelength λ (carrying an energy), which is neither reflected from the device surface nor transmitted through the device, delivers one electron to the external circuit (J_{sc}). The IQE is then defined as IQE $=$ EQE$/(1-$ R$) = f_{abs} \, f_c/(1-$ R$)$. The denominator $(1-$R$)$ determines the amount of incident photons in respect of the reflectance R.

6.5.3 Electron Beam-Induced Current

Connection of electronic devices to an amplifier while under the interrogation of a focussed electron beam of an SEM permits charge collection microscopy, also known as the EBIC technique. The term EBIC is used generally to represent a number of techniques based on the measurement of (a) minority, that is, "injected" carriers generated by energy deposition by a primary beam or (b) the flow of the primary beam current itself, the latter being less common. Since electron beams may be focused and EBIC utilizes minority carriers, the potential of the technique for studying thin-film solar cells with sub-grain scale resolution was recognized early on. Interpretation of all EBIC techniques demands an appreciation of the generation of charge carriers by an electron beam and their collection or else loss in a device. While much can be gained from qualitative interpretation, this is often underpinned by modeling, not least of the electron–hole pair generation function in a structure. A detailed description of this technique, applied to solar cell materials is given in [183, 184]. In Fig. 43, a typical EBIC setup is illustrated.

Another interesting EBIC technique is the cross section or lifetime curve method. In this case, the device under test is cleaved and the beam scanned perpendicular to the layer stack.

The induced current response is measured as the beam traverses the depletion region. Ideally, this takes the subsequent form

$$I_{EBIC} = I_0 \exp\left(\frac{-|x|}{L}\right) \tag{22}$$

Fig. 43 Schematic illustration of an EBIC setup

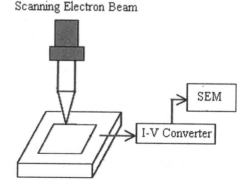

Scanning Electron Beam

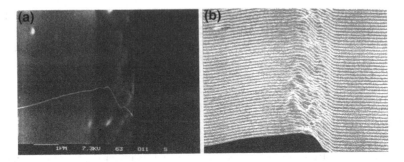

Fig. 44 a SEM image of a cleaved intrinsic a-Si:H on p-type c-Si heterojunction with superposed EBIC line scam, measured with an e-beam energy of 7.3 keV, **b** EBIC map of the same heterojunction, measured with an e-beam energy of 7.3 keV

where x is the distance from the junction, I_0 a constant, and L the minority carrier diffusion length for electrons or holes. Cross-sectional EBIC therefore allows the junction position and homogeneity to be seen and the minority carrier diffusion length to be measured.

An example of the application of this method for the characterization of the a-Si:H/c-Si heterojunction is shown in Fig. 44.

In Fig. 44a, we see the scanning electron microscope (SEM) image of a cleaved intrinsic a-Si:H on p-type c-Si heterojunction with superposed EBIC line scan. An interesting result is that the maximum of the EBIC line scan signal is slightly shifted into the, relatively thick (>1 µm), amorphous silicon layer with respect to the a-Si:H/c-Si interface. In Fig. 44b, the EBIC two-dimensional image of the cleaved heterojunction, measured with an electron beam energy of 7.3 keV, is shown. A part of the non-homogeneity of the signal on the a-Si:H side may be associated to the non-perfect cleaving of the amorphous silicon film.

7 Simulation

There exist three main approaches for the simulation of a-Si:H/c-Si heterojunctions [185, 186, 187]. Most of them are based on one-dimensional models and do not take into account lateral current transport. However, recently, several research groups investigated 2D modeling in heterojunction solar cells using for instance the SENTAURUS device simulator or ATLAS software from Silvaco Inc. [97, 188–190]

In addition, a two-dimensional version of AFORS-HET (see below) is announced but not yet publically available.

The first simulation program [185] developed at the former Hahn–Meitner-Institute (now Helmhotz Zentrum) in Berlin is named AFORS-HET and is based on the finite difference method. The second program developed at the Università la

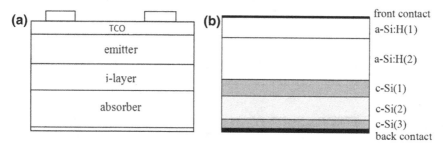

Fig. 45 a Layer sequence of the solar cell, **b** layer sequence, schematically

Sapienza in Rome, Italy, is based on a finite element simulation code (DIFFIN) [186] and permits to take spatial inhomogeneities of the different layers into account. The third one is a development of the University of Oldenburg, Germany [114, 187]. To our knowledge, the first one dominates in practical applications.

To complete the list, there are several other softwares that were developed and adapted or used for silicon heterojunction solar cells, for instance, [191] for AMPS, [192] for SCAPS, and [193] for ASA5.

7.1 AFORS-HET

The principal parameters for the simulation of a (n)a-Si:H/(p)c-Si heterojunction solar cell are as follows:

- Interface state density
- Band offsets
- Properties of the TCO/front contacts
- Properties of the intrinsic buffer layer

Description of the model (cf. Fig. 45):

The emitter layer of the a-Si:H/c-Si solar cell is highly doped. A transparent conductive layer (TCO) is required in order to minimize the electrical and optical losses. A wafer texturization and an i-layer are included.

For the simulation, the amorphous layer is subdivided into two layers: (1) doped a-Si:H and (2) intrinsic a-Si:H.

The crystalline absorber is subdivided into three layers: i) a defect-rich layer of 5 nm, ii) the absorber of 250 μm, and iii) the BSF layer.

One of the goals consists in the comparison of the two structures, nip and pin. In Table 6, the most important material parameters are listed.

In addition, the density of localized states in a-Si:H and the position of the Fermi level (that provides neutrality in a bulk a-Si:H layer) are key parameters that determine the band bending and the open-circuit voltage.

Results:

Table 6 Material parameters used in the simulation

	n/p-type	p/n-type
Band offset	$\Delta E_{cb} = 0.2$ eV	$\Delta E_{vb} = 0.4$ eV
c-Si mobility of the minority charge carriers	$\mu_n^{cSi} = 1{,}000$ cm^2/Vs	$\mu_p^{cSi} = 340$ cm^2/Vs
a-Si:H mobility of the minority charge carriers	$\mu_p^{aSi} = 1$ cm^2/Vs	$\mu_n^{aSi} = 5$ cm^2/Vs

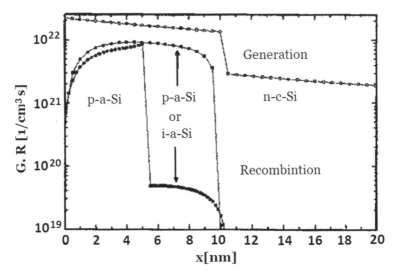

Fig. 46 Simulation of the recombination rate for the short-circuit case in p/n- junctions with 5 nm (p)a-Si:H/5 nm (i)a-Si:H and 10 nm (p)a-Si:H, Ref. [185]

- The diffusion potential of a p/n structure is higher than that of a n/p structure. As a consequence, the open-circuit voltage is higher, too.
- The band offsets constitute a barrier for the minority charge carriers. This leads to a decrease in the fill factor and cannot be compensated by a higher open-circuit voltage.
- Within the i-layer, the recombination rate is distinctly reduced, see Fig. 46.
- The short circuit of the nip-type current is higher than that of pin-type.
- For very thin emitter layers, the front contact induces depletion of the a-Si-layers and enhances the recombination at the junction.

7.2 Comparison with Experiments

Because an example of the AFORS-Het code is shown later in detail (see Sect. 8.2, radiation hardness), an application the DIFFIN simulation of an amorphous silicon on crystalline silicon heterojunction before and after irradiation with 1.7 MeV protons with a flux of 5×10^{12} protons/cm^2 is given first [194]. In the present

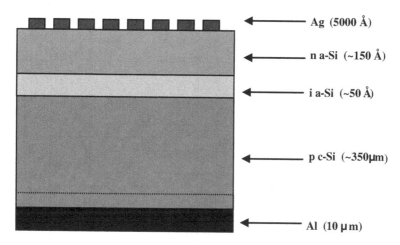

Fig. 47 Schematical drawing of the investigated a-Si:H/c-Si solar cell structure

case, for example, the density of states (DOS) distribution in the intrinsic and in the n-type a-Si:H layers has been fixed as input parameters, and details of the defect distribution in the crystalline silicon substrate have been calculated.

The layer structure of the investigated hydrogenated amorphous silicon/crystalline silicon (a-Si:H/c-Si) hetero solar cell is shown in Fig. 47. The following growth technology has been used: A 500 μm thick ⟨100⟩ oriented, 1 Ω cm p-type crystalline CZ-silicon wafer is exposed to KOH etchant for surface texturing, resulting in an irregular inverted pyramid structure with about 20 μm surface roughness. After the etching, the average wafer thickness is about 350 μm. As next step, the 10-μm-thick back contact metallization is done by aluminum screen printing and subsequent annealing at 750 °C. This annealing step results in a back surface field. After a cleaning procedure, including a dip in 1 % HF for native oxide removal, the 5-nm-thick intrinsic and the 15-nm-thick n-type a-Si:H layers have been deposited. Finally, a 500-nm Ag grid as front contact has been evaporated. The cell area is 1.5 × 1.5 cm. In Fig. 48, the width DIFFIN calculated band structure of the resulting solar cell is shown.

In Fig. 49, the comparison of the current–voltage characteristics under AM 1.5 illumination and in the wavelength dependence of the quantum yield characteristics, as computed by the DIFFIN simulation, are compared to the measured characteristics. A good agreement between measured and simulated curves is found (Table 7).

In Fig. 50, as an important result of the simulation, for example, the recombination rate as a function of the distance from the top metal electrode before and after irradiation is shown. In the inset, one observed enhanced recombination at the intrinsic a-Si:H/c-Si interface. The distribution of the recombination rate within the crystalline silicon layer after irradiation agrees fairly well—at least in the value of the depth value of the degraded layer—with the distribution of the vacancies produced in crystalline silicon as calculated by Monte Carlo calculations using the

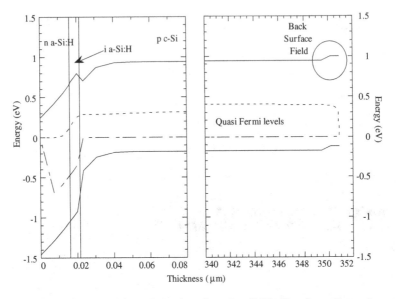

Fig. 48 Simulated band diagram of the investigated a-Si:H/c-Si solar cell structure under illumination (AM 1.5) [194]

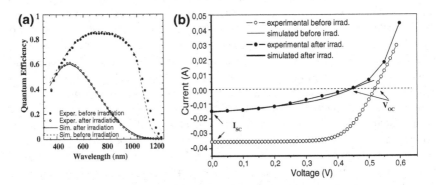

Fig. 49 Comparison of the measured and with DIFFIN-simulated quantum yield spectrum (**a**) and of the current–voltage characteristics (**b**) under AM1.5 conditions before and after irradiation of the hetero solar cell with 5×10^{12} protons/cm^{-2} at 1.7 MeV [194]

Table 7 Solar cell parameters, before and after irradiation with 5×10^{12} protons/cm^2 at 1.7 MeV (from [194])

	Before irradiation	After irradiation
η (%)	12.63	3.4
V_{OC} (V)	0.542	0.449
J_{SC} (mA/cm^2)	35.6	15.4
R_p (Ω)	426	70.4
R_s (Ω)	1.38	20.6
FF (%)	67.7	53.7

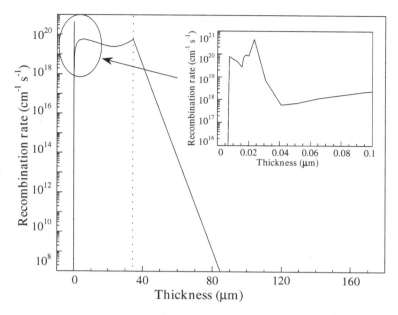

Fig. 50 Distribution of the recombination rate under AM1.5 conditions before (*full line*) and after irradiation (*broken line*) of the hetero solar cell with 5×10^{12} protons cm^{-2} at 1.7 MeV, as obtained by DIFFIN simulation [194]

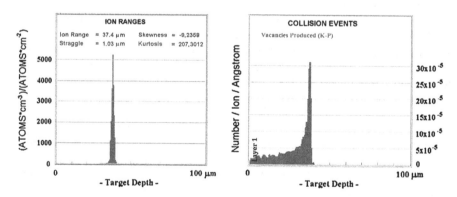

Fig. 51 Monte Carlo simulation with SRIM of (**a**) the distribution of the implanted protons and (**b**) the distribution of the produced vacancies in crystalline silicon after irradiation with 1.7 MeV protons [194]

SRIM code [195] and shown in Fig. 51. It should be mentioned that the simulation results have been found to be rather sensitive to changes in the depth distribution of the recombination rate.

8 Long-Term Stability and Degradation

The long-term stability of solar cells is a field not only of scientific importance but also of large commercial interest. As an example, the German feed in tariff guarantees good government subsidies for 20 years. So it is of vital interest to the consumer (and therefore to the producer) to know the lifetime of his devices. The question is more legitimate since a-Si:H is involved in heterojunction solar cells. The detrimental effects of sunlight to pure a-Si:H cells are well known.

The same argument holds for the radiation experiments. Only the user is different: The space agencies. As a rule of thumb, a device has to survive 1 Mrad intensity during its mission. So any user will subject his devices to a terrestrial load of the same intensity.

8.1 Long-Term Stability

Very few data on long-term stability of heterojunction solar cells have been published in literature. Internal research has been done at the University of Hagen. In a first example (Fig. 52), no degradation after intense UV light source for 1,000 h has been found [196].

A similar investigation was done for an amorphous emitter by the same author [196]. After 60-h irradiation at five suns, a negligible efficiency decrease from 12.4 to 11.8 % was observed.

A more recent investigation on long-term stability of heterojunction solar cells is given in Ref. [197]. The main emphasis is laid on the surface recombination velocity, s_{eff}. Light soaking at AM1.5 and 50 °C and dark degradation (ambient atmosphere in a drawer at about 22 °C) both for 1,000 h is applied. For the technically interesting case of 5-nm-passivating a-Si:H layers, s_{eff} increases by a factor of 4 for dark degradation and a factor of 2 for light soaking.

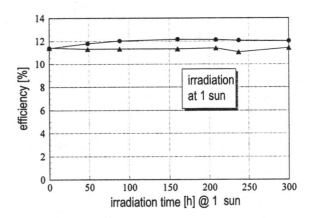

Fig. 52 Efficiency versus irradiation time at one-sun intensity. The circles represent an n-type microcrystalline emitter while the triangles represent an n-type emitter and an intrinsic buffer layer

Another hint for the long-term stability is given in [31]. The supplier of HIT solar modules guarantees a degradation of less than 20 % within 20 years. This is the usual degradation valid for most types of solar cells.

8.2 Radiation Hardness

8.2.1 Proton Irradiation

Intensive measurements on radiation hardness data of heterojunction solar cells are carried out in [198]. These data are mainly required for the purpose of space missions. In contrast to the 1950s and 1960s of the last century when only monocrystalline solar cells were known a-Si:H/c-Si heterojunction solar cells now are interesting candidates. The reason is their simple, low-temperature, and low-budget technology, and the higher efficiency at high temperatures [32] compared to conventional cells.

The research concentrates on proton irradiation because these particles are widely used and the obtained numbers can easily be converted to those of other energies or other radiation sources [199]. For instance, it is customary to normalize proton damage constants of a given energy to those of the energy of 10 MeV. Similarly, the damage constants after electron, gamma, or neutron irradiation are converted to that of proton irradiation.

Several optical and electrical parameters of a-Si:H/c-Si heterojunction solar cells are measured before and after a proton irradiation of varying irradiation doses (cf. Table 8). Representing proton damage in the solar cell by an inhomogeneous defect distribution that is based on a calculated damage profile using the TRIM software [195], the experimental data can be well reproduced within the AFORS-HET simulations.

The solar cells are irradiated with protons of energies of 0.8, 1.7 MeV, and 4 MeV. Samples with irradiation doses in the range from 5×10^{10} to 5×10^{12} cm^{-2} and non-irradiated control samples are investigated. Several optical and electrical parameters are measured among them are the internal quantum efficiency, IQE, versus wavelength, λ (Fig. 53), and characteristic electrical parameters (Table 8; for the sake of simplicity, the results of the later simulation in this table are anticipated) are shown.

It is clearly visible that the long-wavelength tail decreases with increasing proton dose. This behavior is identical to that of fully crystalline cells [200, 201] and indicates that most of the damage occurs in the crystalline silicon. This behavior might be expected when looking at the displacement distribution caused by the protons (Fig. 54). The relatively low-energy deposition in the near-junction regions (later labeled regions A) provokes minor permanent damage since this region is just passed by the protons with high energies. The same behavior can be expected for neutrons for which a confirmation is found in [202].

Table 8 Experimental and simulated values of I_{SC}, V_{OC}, FF, and η for various irradiation doses at 1 MeV [198]

Dose (p/cm^2)	Experiment				Simulation			
	J_{SC} (mA/cm^2)	V_{OC} (mV)	FF (%)	η (%)	J_{SC} (mA/cm^2)	V_{OC} (mV)	FF (%)	η (%)
Control	29.3	613	79.9	13.7	29.3	613	79.9	13.7
6.5×10^{10}	25.4	567	71.5	10.3	25.4	567	71.5	10.3
2×10^{11}	24.2	554	66.7	8.9	24.2	554	66.7	8.98
1×10^{12}	19.0	527	58.9	5.9	19.0	527	58.9	5.91
5×10^{12}	16.7	501	67.8	5.6	16.7	501	67.8	5.69

Fig. 53 Measured internal quantum efficiencies of solar cells irradiated with different doses of 1.7 MeV protons and one non-irradiated control sample [198]

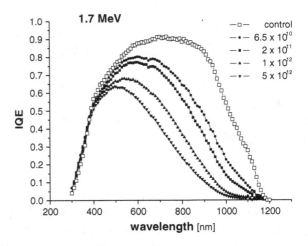

Fig. 54 Simulated damage distributions for protons of 0.5, 1.7 and 4 MeV. 10^5 particles are used as parameter for the simulation done with TRIM software [198]

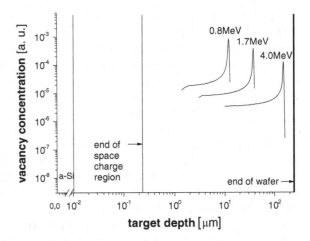

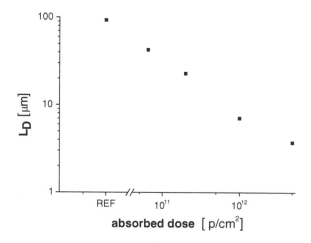

Fig. 55 Diffusion length, L_D, versus dose, ϕ, of the samples irradiated at 1.7 MeV with doses of 6.5×10^{10}, 2×10^{11}, 1×10^{12}, 5×10^{12} and a non-irradiated control sample. L_D is estimated from the IQE values (Fig. 53) in the long-wavelength range (880–970 nm) [198]

The measured quantum efficiency is converted to the bulk diffusion length, L_D, according to a commonly applied analysis [203, 204] using a plot of the inverse quantum efficiency $IQE^{-1}(\alpha-1)$ versus the inverse of the absorption coefficient, $\alpha^{-1}(\lambda)$. The effective diffusion length is derived from the slope of this curve. In Fig. 55, L_D is plotted as a function of the applied dose.

This plot gives access to a figure, which is a measure of the radiation hardness of a solar cell, namely the damage constant, k_L. Let us start with a plausible assumption, namely that the number, N_{tr}, of electrical active defects (e.g., displacement, interstitials, complexes formed thereof) is proportional to the applied dose, ϕ, so that $N_{tr} \propto \phi$ applies. From Shockley–Read–Hall statistics, it is known that $N_{tr} \propto 1/\tau$, τ being the bulk lifetime. Since $L_D = \sqrt{(D \tau)}$, we end up with

$$\frac{1}{L_D^2} = \frac{1}{L_{D0}^2} + k_L \Phi \tag{23}$$

k_L, originally just a proportionality constant, stands for number of traps induced by the dose. The lower the k_L, the better the radiation hardness of the cell. L_{D0} is the diffusion length prior to irradiation and reflects the number of traps introduced during the production of the wafer. A typical $1/L_D^2$ versus Φ plots is seen in Fig. 56.

The deduction of k_L is repeated for other energies. In Fig. 57, the a-Si:H / c-Si-Si results are compared with damage constants from earlier work, that is, for fully crystalline silicon cells [205–207]. It is evident that the heterojunction solar cell damage data are amidst the data band of the crystalline cells or at its lower end. This is another indication that the substrate is responsible for the degradation and that the amorphous silicon (a-Si:H) is little involved.

In Fig. 58, this comparison is repeated in more detail. A characteristic peak emerges at about 1.5 MeV. To explain this peak, the substrate is subdivided into three regions, namely left (region A) and right (region B) of the implantation maximum, and near the implantation maximum, say one half-width right and left

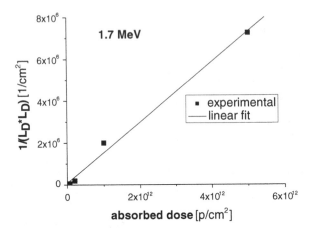

Fig. 56 $1/L_D^2$ versus Φ for the samples irradiated at 1.7 MeV with doses of 6.5×10^{10}, 2×10^{11}, 1×10^{12}, 5×10^{12} cm^{-2} and a non-irradiated control sample. The slope delivers the damage constant, $k_L = 1.4 \times 10^{-6}$ [198]

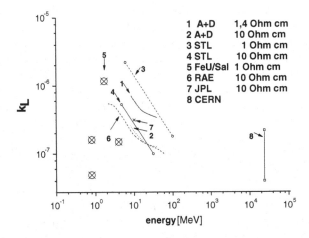

Fig. 57 Damage constants versus proton energy from literature and from [198]. The data labeled A + D are measured by Anspaugh and Downing ([206], Fig. 6), the data labeled STL (Space Technology Laboratories, Redondo Beach, CA) and RAE (Royal Aircraft Establishment, Farnborough, UK) are included therein, and the data labeled CERN (Centre Européen pour la Recherche Nucléaire) are measured on npn transistors and taken from [207], Table 7.2. The data labeled JPL are taken from [205], Fig. 3.5

of it (region C). Now, a result of the simulation is anticipated: The degradation of the solar cell is mainly controlled by region A. Thus, the product $P = N_{tr} \times \text{length}(A)$ is taken as a measure for k_L. Figure 54 yields $P = 5 \times 10^{-4}$ (0.8 MeV), 6×10^{-4} (1.7 MeV), and 4×10^{-4} (4 MeV) in the arbitrary units

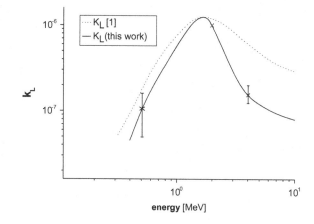

Fig. 58 Damage constants versus proton energy. *Dashed line* Saturation current data from [205], Fig. 4.3. *Full line* [198]. The line is only a guide to the eyes

given by the axes. This is a tentative explanation for the occurrence of a k_L maximum at 1.5 MeV.

The observed peak is in good agreement with an earlier measurement of crystalline damage coefficients obtained in [205] via the saturation current of a crystalline solar cell. Basically, a curve with identical features is presented there. In both cases, the degradation constant shows a maximum around 1.5 MeV. Note that the energy of 6 MeV is sufficient so that the protons just pass the wafer and little permanent displacement damage occurs.

In order to interpret the effect of proton damage with the help of numerical computer simulation, in a first step, the IV curves in the case of non-irradiated samples are reproduced. The following measured data are used as a direct input for the AFORS-HET simulation: The layer sequence, that is, using a 10-nm-thick layer of n-doped a-Si:H on top of a 230-μm-thick p-doped silicon wafer, c-Si, with 1 Ω cm resistivity, the measured reflection of the cell, $R(\lambda)$, and the measured absorption loss within the ITO layer, $A(\lambda)$ (obtained by Ulbricht sphere measurements). The trap density of a single midgap defect within the c-Si bandgap is adjusted, so that it reflects the stated mean effective diffusion length (e.g., $N_{tr} = 2.6 \times 10^{12}$ cm^{-3} corresponds to $L_{eff} = 119$ μs for the non-irradiated solar cells). In order to model the defect distribution of the a-Si:H layer, two exponential distributions toward the band edges (Urbach tails) and two Gaussian profiles (dangling bond distribution and correlation energy) within the a-Si:H bandgap are assumed. All these distributions are constant in space. The magnitude of the following a-Si:H parameters have been obtained experimentally: The position of the Fermi energy within a 10-nm-thin a-Si:H(n) layer deposited on a c-Si(p) wafer of 1 Ω cm resistivity, that is, the distance of the Fermi energy to the valence band, $E_F - E_V = 1.3$ eV, together with the Urbach energy of the valence bandtail, $E_{uv} = 73$ meV and the Gaussian dangling bond distribution, $E_{db} = 500$ meV, $\sigma_{db} = 230$ meV, $N_{db} = 2.33 \times 10^{19}$ cm^{-3}, has been measured for similar a-Si:H/ c-Si solar cells with the help of photoelectron spectroscopy [208]. The Urbach energy of the conduction band, $E_{uc} = 106$ meV, and the correlation energy,

Fig. 59 Measured and AFORS-HET-calculated IV curves for non-irradiated samples [198], cell area of 1 cm^2

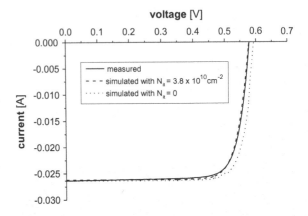

$E_{cor} = 200$ meV, has been guessed. Furthermore, in Ref. [208], a 1.74-eV a-Si:H bandgap has been measured and the valence band offset ΔE_V has been determined to be in the range $\Delta E_V = (450 \pm 50)$ meV. The doping efficiency of a-Si:H was adjusted in a way in order to reflect the measured Fermi energy position in a-Si:H.

As seen from the above paragraph, only few parameters are left to be varied in the simulation. Among these, there are mainly the interface state and bulk trap densities, the band mismatch and the surface recombination velocities. The simulations have concentrated on the interface and bulk state densities because these ones can be expected to be most decisive to the IV and SR parameters. It is well known that a non-vanishing interface state density between the amorphous and the crystalline silicon layers critically influences the solar cell performance, that is, the open-circuit voltage of an a-Si:H/c-Si solar cell [185]. In order to simulate the measured IV curves in case of non-irradiated samples, a non-vanishing constant interface density distribution N_{it} has been introduced.

In Fig. 59, the result of the fitting is shown with an energy-integrated interface density of $N_{it} = 3.8 \times 10^{10}$ cm^{-2}. For comparison, the IV curve of a sample with $N_{it} = 0$ is also shown. No way was found to obtain a good fit for V_{oc} based on a vanishing interface density using the measured parameters cited above. Using this parameter set, not only the IV curve but also the corresponding spectral response of the non-irradiated sample are already sufficiently reproduced, cf. Fig. 60. Apart from a minor displacement of about 15 nm, the simulated and experimental curves are almost identical. In order to properly adjust this shift, one should adopt the recombination velocity at the rear contact and optimize the discretization of the c-Si wafer.

In a second step, the influence of the proton irradiation dose on the spectral response is examined. For this purpose, the damage profile, which has been calculated with the TRIM software, is approximated by three different regions of constant c-Si bulk trap densities, N_{tr}. The wafer is subdivided into three regions, namely left (region A) and right (region B) of the implantation maximum, and near the implantation maximum, say one half-width right and left of it (region C), cf. Fig. 61. Region B, which is not affected by the irradiation, bears the native defect

Fig. 60 SR for a 4 MeV, 2×10^{11} cm^{-2} irradiated and control sample as measured (*circles and stars*) and simulated (*open and full*). Trap densities are 2.2×10^{12} cm^{-3} (*control sample*) and 2.5×10^{13} cm^{-3} (*irradiated sample*) [198]

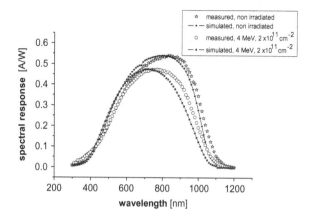

Fig. 61 The calculated trap distribution (simulation with TRIM software) is replaced by a simplified three-layer trap distribution used for the AFORS-HET simulation. *Note* that all damage is within a region of front to 50 % of the sample thickness [198]

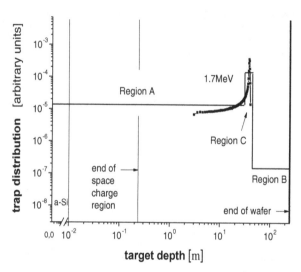

density, which is determined from the simulated sample without proton irradiation. The density of region C has little impact on the simulation results; for simplicity, a density 10 times of that of region A is assumed. Thus, the density of region A essentially controls the quality of the fit. Based on this trap profile due to proton irradiation, the spectral response (SR) for a 4 MeV, 2×10^{11} cm^{-2} implanted sample is simulated.

The result of the SR fit together with the non-irradiated reference is given in Fig. 60. The effect of proton irradiation on the spectral response is clearly reproduced within the AFORS-HET simulations: The experimentally observed decrease in the maximum and of the high-wavelength response together with no change in the low-wavelength region is adequately observed by numerical simulation. In the simulation, the trap density is varied until the simulated SR maxima are identical to the experimental ones. The trap fitting procedure is applied to

Fig. 62 SR for a 1.7 MeV, 5 × 10¹² cm⁻² irradiated sample as measured (*open circles*) and simulated (*full circles*) [198]

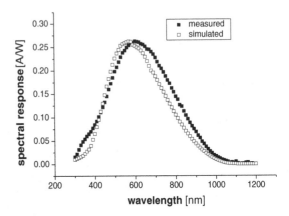

Fig. 63 Trap density of region A obtained from SR simulation versus applied dose. 1.7 MeV, doses 0, 6.5 × 10¹⁰, 2.0 × 10¹¹, 1.0 × 10¹², and 5.0 × 10¹² p/cm² [198]

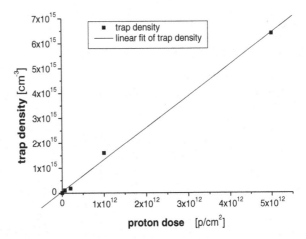

samples implanted with 1.7 MeV and various doses. An example of simulated and experimental SR can be seen in Fig. 62.

In Fig. 63, the bulk trap density of region A is plotted versus the applied dose. From this figure, we conclude that the basic assumption, namely $N_{tr} \propto \phi$ is entirely justified. The trap densities in region A resulting from the fitting procedure are gathered in Table 9.

Influence of the insertion of a thin intrinsic buffer layer on the degradation kinetics after proton irradiation:

While so far proton degradation results, regarding n-type a-Si:H on p-type c-Si heterojunction solar cells without intrinsic buffer layer have been reported, now we want to demonstrate the particular influence of the insertion of an intrinsic a-Si:H interface buffer layer on the device degradation during proton irradiation. The structure of the two different solar cells that will be compared in this chapter is schematically shown in Fig. 64. Identical structures have been used, with an addition of a 5-nm-thick intrinsic buffer layer in the case of the second heterodiode.

Table 9 Substrate trap density of the non-irradiated samples and trap densities of region A, C, and B of samples irradiated at 1.7 MeV [198]

Doses (p/cm^2)	Trap densities for 1.7 MeV		
	Region A (cm^{-3})	Region C (cm^{-3})	Region B (cm^{-3})
0	2.6×10^{12}	2.6×10^{12}	2.6×10^{12}
6.5×10^{10}	1.1×10^{14}	1.1×10^{15}	2.6×10^{12}
2.0×10^{11}	1.8×10^{14}	1.8×10^{15}	2.6×10^{12}
1.0×10^{12}	1.6×10^{15}	1.6×10^{16}	2.6×10^{12}
5.0×10^{12}	6.4×10^{15}	6.4×10^{16}	2.6×10^{12}

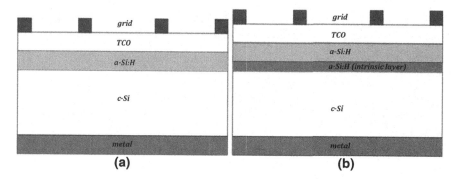

Fig. 64 The structure of a heterojunction solar cell without (**a**) and with (**b**) intrinsic layer

For the irradiation, again 1.7 MeV protons have been used with fluences ranging from 5×10^{10} to 5×10^{12} protons/cm^2. In Fig. 65, the comparison of the quantum yield spectra before and after degradation for both type of heterostructures is shown.

Both heterojunctions show quite a similar degradation kinetics in the wavelength range above 600 nm, but a remarkable difference in the short wavelength ranges from 400 to 600 nm. For all solar cells with intrinsic layer, the quantum yield increases at smaller wavelengths after proton irradiation in a very similar manner. This effect is absent for the solar cells without intrinsic layer. We conclude that it is strongly correlated to an irradiation-induced modification of the intrinsic layer. We did observe important modification of intrinsic a-Si:H film electrical properties, as determined by temperature-dependent dark current conductivity measurements [209]. In Fig. 66, we see the irradiation dose dependence of the effective charge carrier diffusion length, L_{eff}, as obtained by a fit of these characteristics in the long-wavelength range (procedure as described in [194]).

L_{eff} decreases monotonically from values of about 160 μm before irradiation down to values of about 6 μm for the highest dose of 5×10^{12} protons/cm^2 for both types of heterojunctions. As expected, there are no significant differences with respect to the insertion of the thin intrinsic layer, because changes in L_{eff} are mainly due to lifetime changes in the crystalline silicon absorber layer.

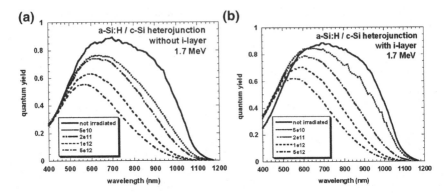

Fig. 65 Quantum yield spectra of two different a-Si:H/c-Si heterostructure solar cells without (**a**) and with (**b**) i-layer insertion, before and after irradiation with 1.7 MeV protons

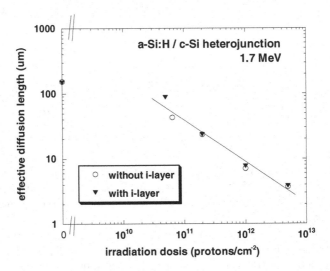

Fig. 66 Comparison of the effective minority carrier diffusion length within the crystalline silicon as a function of the proton irradiation doses for two types of a-Si:H/c-Si heterojunction solar cells with (*triangles*) and without (*open circles*) thin intrinsic buffer layer

In Fig. 67, the normalized emission spectra of a conventional c-Si solar cell and an a-Si:H/c-Si heterojunction solar cell, both forward biased with a current density of 30 mA/cm^2, have been compared. Very similar spectra have been measured, both due to the band-to-band crystalline silicon emission. The heterojunction emission is slightly broader, because of the "window" effect of the higher bandgap amorphous emitter, resulting in a reduced self-absorption of the EL signal. For the full width half maximum (FWHM) of the emission spectrum, a value of 95 nm has been measured for the homojunction solar cells, that is slightly lower than the value of 110 nm, measured for the heterojunction silicon solar cells. The influence

Fig. 67 Comparison between the room temperature electroluminescence spectra of a a-Si:H/c-Si heterojunction and of a conventional crystalline silicon homojunction solar cell

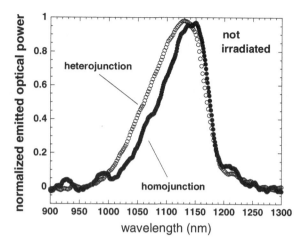

Fig. 68 Influence of high-energy proton irradiation on the room temperature electroluminescence spectrum of an a-Si:H/c-Si heterojunction

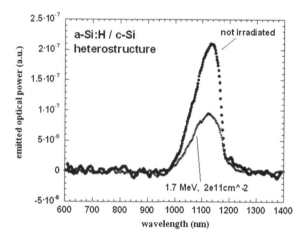

of the proton irradiation on the EL signal of a-Si:H/c-Si solar cells is shown in Fig. 68. We observe, as mentioned earlier, the strong decrease in the electroluminescence signal without change in the emission spectral characteristics after proton irradiation.

Plotting the electroluminescence intensity versus the applied bias current for a series of samples, that have been irradiated at 1.7 MeV with different fluences of protons (see Fig. 69), we observe above a current threshold value an almost linear increase in the optical power with increasing current.

We can define a differential electroluminescence quantum efficiency in the linear region that is given by the optical power change, divided by the respective bias current change.

In Fig. 70, these electroluminescence quantum efficiency values of a series of solar cells, with and without intrinsic buffer layer, are plotted as a function of the proton fluence. It can be clearly seen that the insertion of the thin intrinsic layer

Fig. 69 Influence of high-
energy proton irradiation (4
different fluences) on the
injected current versus
emitted optical power
characteristics at room
temperature of a a-Si:H/c-Si
heterojunction solar cell

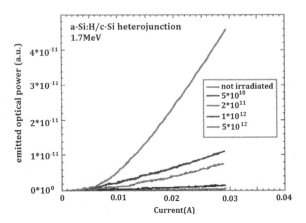

Fig. 70 Influence of the
proton irradiation (1.7 MeV)
fluence on the differential
electroluminescence
efficiency of forward-biased
a-Si:H/c-Si heterojunctions
with or without intrinsic
buffer layer at the
heterointerface

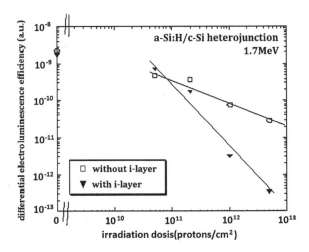

results in a faster degradation with increasing proton doses regarding the electroluminescence efficiency. This results may reflect the much stronger changes in the intrinsic layer properties, as compared to the doped layer properties upon irradiation, because of the lower initial deep defect density in the i-layer.

8.2.2 Electron Irradiation

The photocurrent enhancement in the spectral range below 600 nm after proton irradiation of a heterojunction solar cell with intrinsic a-Si:H buffer layer, shown in Fig. 70, is one of the rare examples of a partial device improvement of the optoelectronic properties for silicon-based devices. A similar observation has been found in the case of the electron irradiation at 1 MeV of the same type of solar cell.

Fig. 71 Current–voltage characteristics under AM1.5 illumination of a-Si:H/c-Si heterojunction solar cells before and after irradiation with 1 MeV electrons for different fluences

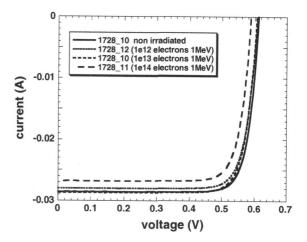

Fig. 72 Dependence of different electron fluences on the I_{SC} and V_{OC} changes in a-Si:H/c-Si heterojunction solar cells, measured under AM1.5 illumination conditions

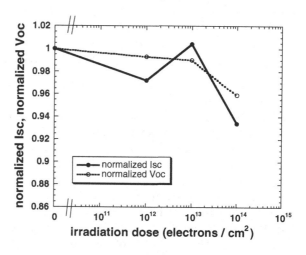

In the current–voltage characteristics under illumination (Fig. 71), we observe the maximum of the photocurrent at a fluence of 1×10^{13} electrons/cm^2. This can be more clearly observed in Fig. 72, where the short-circuit current and the open-circuit voltage values have been plotted as a function of the electron fluence. The maximum value for the short-circuit current is found at an intermediate electron fluence value. The observed photocurrent enhancement is probably related to the easier electronic transport through the spike in the conduction band at the heterointerface between the amorphous and crystalline silicon with increasing irradiation dose due to a defect-assisted tunneling mechanism.

Table 10 State of the art

Producer/Research Group	Base	Area [cm^2]	V_{oc} [mV]	Eff [%]	Ref.
Sanyo (Japan)	n-type CZ	100	745	23.0	[210]
Roth&Rau (Switzerland)	n-type CZ	4	735	21.9	[211]
HEMRI (Korea)		220	721	21.1	[212]
Kaneka and IMEC (Japan and Belgium)	n-type	156		>21.0	[213]
CEA-INES (France)	n-type FZ	105	732	21.0	[214]
IMT and EPFL (Switzerland)	n-type FZ	4	725	21.0	[215]
FernUniversität Hagen (Germany)	n-type FZ	1	675	19.7	[216]
HMI (Germany)	n-type FZ	1	675	19.3	[217]
NREL (USA)	p-type FZ	0.9	678	19.3	[218]
TiTech (Japan)	p-type FZ	0.8	680	19.1	[219]

9 State of the Art

The state of the art can be demonstrated best by means of the Table 10. Here, the results of the main companies and academic institutions dealing with a-Si:H/c-Si heterojunction solar cells have been gathered.

The list is not complete. In order to avoid overloading, just those institutions are shown which have overcome the 19 % limit. The following abbreviations were used:

HEMRI for the Hyundai Electro-Mechanical Research Institute of Hyundai Heavy Industries in Yongin (Korea),
CEA-INES for the Commissariat à l'Energie Atomique—Institute National de l'Energie Solaire in Annnecy (France),
IMEC for the Interuniversity Microelectronique Center in Leuven (Belgium),
IMT-EPFL for the Institute de Microtechnique in Neuchatel (Switzerland) of the Ecole Polytechnique Fédérale de Lausanne,
HMI for the (former) Hahn-Meitner-Institute in Berlin (Germany), now Helmholtz Center of Materials and Energy (HZB),
NREL for the National Renewable Energy Laboratory in Boulder (USA),
TiTech for the Tokyo Institute of Technology (Japan).

Sanyo is the unbeaten leader in this race. The decisive parameter for its superiority probably is the high V_{oc} of 745 mV. It is known that V_{oc} is mainly controlled by the passivation quality. It might also be degraded by local shorts, for example, when pyramids of the texturization penetrate the upper layers. At this point, a large field of research is still open.

10 Silicon-Based Heterojunction Solar Cells in China

In China, the investigation of the silicon-based heterojunction solar cells was started in the 1980s. The first report was published by Jiang et al. [220] of the Beijing Institute of Solar Energy in 1982. The structure of the cell was Ag/ITO/

(n)c-Si/Ag. All the films were finished by thermal evaporation. The efficiency of the cell was about 5.4 % (100 mW/cm^2, halogen lamp). In the same year, another cell with a SnO$_2$/(n)c-Si structure on single-crystalline silicon was reported by the same group [221]. In 1987, a SnO$_2$/SiO$_x$/polySi heterojunction cell which reached an efficiency of 10 % (AM1.5) on the light receiving area was produced by Liu et al. The SiO$_x$ layer was made by thermal oxidation, and the SnO$_2$ layer by thermal decomposition. The effects of the grain size of the polysilicon and the thickness of the SiO$_x$ layer on the properties of the cells were analyzed [222].

In the end of the 20th century and in the beginning of the 21st century, more groups focused on silicon-based solar cells with an a-Si:H/c-Si structure, for the upcoming energy crisis and the greatly attractive efficiency of the HIT cells made by Sanyo. In the following, the works of those will be introduced as experimental study and theoretical simulation. The active groups in this area will be introduced.

10.1 Introduction of the Active Groups in this Area in China

Probably, the knowledge about some groups is lost, so we apologize if some information is incomplete or even missing.

10.1.1 Graduate University of the Chinese Academy of Sciences

The first institute to mention is the Graduate University of the Chinese Academy of Sciences (GUCAS), in Beijing. The major members are Fengzhen Liu, Meifang Zhu, and Yuqin Zhou. They mainly focus on Hot-Wire CVD method to produce the a-Si:H/c-Si heterojunction solar cells. They try to improve the interface properties and improve the texturization pyramids using acid rounding or some new chemical recipes, such as tetramethylammonium hydroxide (TMAH), NaOH, and NaClO. To the author's knowledge, the highest efficiency they have achieved is 17–18 %.

10.1.2 Institute of Electrical Engineering, Chinese Academy of Sciences

The second group belongs to the Institute of Electrical Engineering of the Chinese Academy of Sciences (IEE). The major members of this group are Wenjing Wang, Lei Zhao, Chunlan Zhou, and Hongwei Diao. The essential method they use to fabricate the a-Si:H/c-Si heterojunction is RF-PECVD. They also try to improve the texturization of the cells, aimed to make larger size pyramids than those for the conventional c-Si cells. Besides this, they have done efforts to analyze the effect of the layers of the cells on efficiency, open-circuit voltage, short-circuit current, and fill factor of the device. They have obtained the support for developing a pilot line together with Shanghai Jiao Tong University and Shanghai Chaori Solar Energy Science and Technology Company.

10.1.3 Nanchang University

The major members of this group are Lang Zhou, Wolfgang. R. Fahrner, and Haibin Huang. This is a new group which started in 2010. They focus on the technique of the a-Si:H/c-Si heterojunction on solar cell-grade c-Si. They have got an average lifetime better than 800 μs (measured by Semilab WT-2000PV) for a 4 by 4 cm^2 solar cell-grade n-type c-Si wafer, bifacially passivated by PECVD. They have made a complete device by PECVD, magnetron sputtering, and chemical etching.

10.1.4 Shanghai Institute of Microsystem and Information Technology

*The Shanghai Institute of Microsystem and Information Technolo*gy (SIMIT) is one of the institutes of the Chinese Academy of Sciences. The group leader for silicon-based heterojunction solar cell in this institute is Zhenxin Liu. It is also a new group (started in 2010). It has obtained financial support for research on heterojunction c-Si-based solar cells, in cooperation with Trina Solar Company (in Changzhou city). The efficiency of their a-Si:H/c-Si cells was about 19 % (not confirmed).

10.1.5 Other Groups

There are also some others groups, such as the South China University of Technology, in Guangzhou city, the Ningbo University, the Huazhong University of Science and Technology in Wuhan city, and the Jingdezhen Ceramic Institute. They have done some work on simulation. The Hebei University of Technology in Tianjin city, the Shanghai Institute of Space Power Sources, and the Jiangnan University in Wuxi city, have performed experimental studies on silicon-based heterojunction solar cells.

10.2 Experimental Studies

10.2.1 Texturization and Cleaning

Techniques for texturization and cleaning of a-Si:H/c-Si heterojunction photovoltaic cells are quite different compared with those of the conventional cells because the pn junction for the conventional cell is formed by diffusion while the junction for the heterojunction cell is made by stacking new thin layers on the surfaces of the c-Si substrate.

Texturization

Two directions to improve the texturization are followed by the Chinese research groups. One is producing larger pyramids and rounding them to reduce the energy singularity (sharp corners and borders). The other one is producing smaller size pyramids in order to render the surface more uniform.

In Ref. [223], HNO_3, CH_3COOH, and HF solutions were used to polish the texturized pyramids and remove the metal contamination left on the surface. The results showed that an optimized etching time of 10–15 s was effective for reducing the stress in the i-layer. Therefore, the deposition coverage and the contact of each layer could be improved which is beneficial for the performance of the solar cells (Table 11, Fig. 73).

GUCAS showed in 2010 that a c-Si wafer texturized by TMAH had a better lifetime than a wafer texturized by alkaline solution [224] because there is no metal contamination in TMAH. IEE have successfully obtained pyramids larger than 10 μm by TMAH and IPA solution, as shown in Fig. 74 [225]. The sample with 60-min etching time exhibited the lowest reflectivity, 9.32 %. They also tried rounding the pyramids with chemical solutions.

The Nanchang University tried to use a KOH/IPA solution to get larger size pyramids because this solution is much cheaper than TMAH. The group has produced pyramids of about 10 μm (as shown in Fig. 75) by adjusting the compositions and the temperature of the solution [226].

IEE used ultrasonic-enhanced NaClO and TMAH solutions to make small size pyramids, about 4 μm large [227]. The Graduate University of the Chinese Academy of Sciences employed Na_3PO_4 to fabricate dense and small size pyramids [228]. I_{sc} and V_{oc} of the cell texturized by Na_3PO_4 are better than of those texturized by NaOH, as shown in Fig. 76, which is attributed to the advantage of lower surface reflectivity and easier film covering by Na_3PO_4 texturization.

Table 11 Average angle of the top of pyramid peaks and elemental composition on the wafer surface with different chemical polishing times (t_{cp}) using HNO_3, CH_3COOH, and HF solutions [223]

t_{cp} (s)	0	5	10	15	20	30	60	120
Average angle (°)	75.52	76.58	78.65	80.31	82.04	82.85	86.35	92.34
O (%)	1.90	3.38	5.21	2.92	1.49	3.48	2.37	3.17
Na (%)	0.63	0.49	0.73	0.73	0.66	0.84	0.66	0.57
Si (%)	97.47	96.13	94.06	96.35	97.85	95.68	96.97	96.36

Fig. 73 Solar cell properties, such as V_{oc}, I_{sc}, FF, and ς, as a function of the chemical etching time [223]

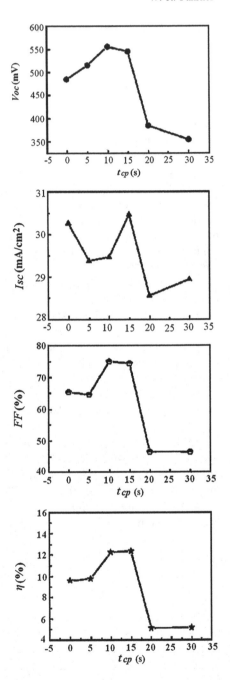

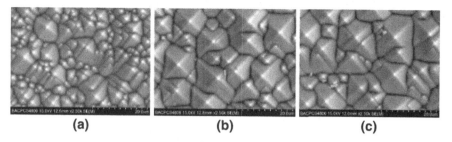

(a) **(b)** **(c)**

Fig. 74 SEM of pyramids with different texturization time: **a** 30 min, **b** 60 min, **c** 80 min [225]

Fig. 75 Pyramids about 10 μm obtained with a KOH/ IPA solution by Nanchang University [226]

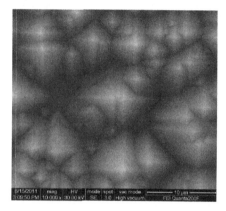

Cleaning

The cleaning methods of the c-Si wafer for the heterojunction solar cells are based on the RCA cleaning recipes; HF is used to remove the natural oxide layer at the end of the cleaning process. [224, 226, 228]. The Nanchang University found that the pre-cleaning step (with or without a RCA step) before texturization has a large effect on the pyramid morphology [226]. An unsatisfactory texturization on a 4×4 cm^2 c-Si wafer is shown in Fig. 77, which is with a RCA cleaning step before the texturing.

As the last step of the wafer cleaning process, normally dipping the wafer into HF solutions for a short time is used to remove the natural oxidation layers and to passivate the surface with H bonds. The H bond passivation is good for preventing the natural oxidation again, but it could not last for a long time. This limits the length of the time allowed between the HF dipping and the subsequent process. This length of the time was studied by IEE. The effects of the immersion time in a HF solution and the exposure time in the air on the effective lifetime, τ_{meas}, were measured as shown in Fig. 78 [229].

Shuzhen Wang and Zhengrong Shi studied the effect of the cleaning solution before HF dipping on the lifetime, as shown in Table 12. It is found that a solution of concentrated sulfuric acid and H_2O_2 is the best one. According to their opinion,

Fig. 76 Light I–V curves of
the HIT cells with **a** NaOH
texturization, **b** Na_3PO_4
texturization, **c** Dark I–V
curves of the two samples
[228]

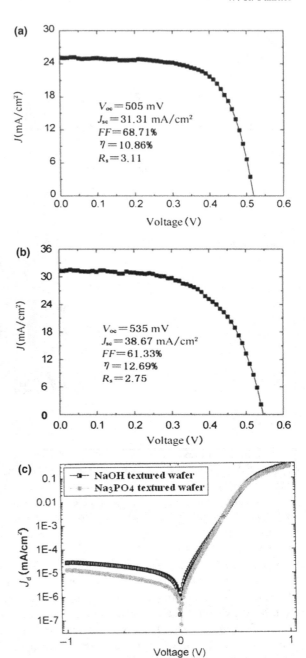

(a)

$V_{oc} = 505 \, mV$
$J_{sc} = 31.31 \, mA/cm^2$
$FF = 68.71\%$
$\eta = 10.86\%$
$R_s = 3.11$

(b)

$V_{oc} = 535 \, mV$
$J_{sc} = 38.67 \, mA/cm^2$
$FF = 61.33\%$
$\eta = 12.69\%$
$R_s = 2.75$

(c)

NaOH textured wafer
Na3PO4 textured wafer

the reason is that this solution reduces the roughness and surface defect states
better than any alternatives, as shown in Table 12 [230]. Schematic cross section
of a silicon wafer surface treated with different solutions is shown in Fig. 79.

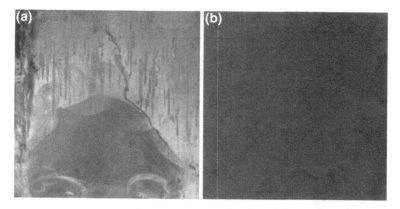

Fig. 77 a Morphology of an unsatisfactory texturization on a 4 × 4 cm^2 c-Si wafer. RCA cleaning is applied before texturization; **b** morphology of a satisfactorily texturized wafer without RCA cleaning before texturization [226]

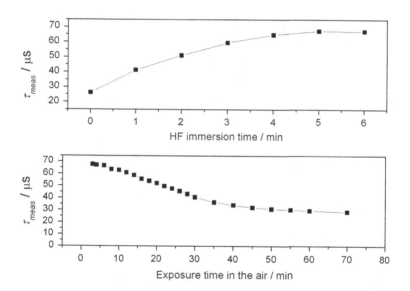

Fig. 78 The effective lifetime (τ_{meas}) of the texturized (p)c-Si wafer versus immersion time in HF of 1 %, and versus exposure time in air after immersion in the solution [229]

10.2.2 Passivation

The best passivation has been obtained by the Nanchang University. The passivation layer is an a-Si:H film covered by PECVD. The average lifetime of the wafer after bifacial passivation is about 880 μs (as shown in Fig. 80), measured by the Semilab WT-2000PV μ-PCD method. The wafer is of solar cell-grade n-type monocrystalline silicon and its size is 4 × 4 cm^2.

Table 12 Lifetime of minority carriers using different cleaning methods [230]

Samples	Chemical solution	Temperature (°C)	Time (min)	Lifetime (µs)
pe1	1 %HF + 3 %H$_2$O$_2$	RT	1	77.803
pe2	1 %HF	RT	1	72.646
pe3	40 %NH$_4$F	RT	2	48.288
pe4	H$_2$SO$_4$ (concentrated) + H$_2$O$_2$(1:1)	120	12	107.880
	1 %HF	RT	2	

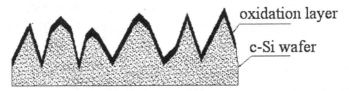

(a) Schematic diagram of silicon wafer surface after RCA cleaning

(b) Schematic diagram of silicon wafer surface after oxidizing in the mixture of undiluted H$_2$SO$_4$ and H$_2$O$_2$

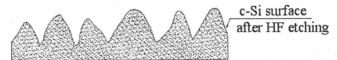

(c) Schematic diagram of silicon wafer surface after etching in the solution of diluted HF

Fig. 79 Schematic cross section of a silicon wafer surface treated with different solutions [230]

Nanchang University also has studied the effect of the post-annealing on the lifetime of a-Si:H-passivated and of a-SiO$_x$:H-passivated wafers [232, 233]. They found that 300 °C and 60 min was the best choice for the improvement of the lifetime for both kinds of wafers, as shown in Fig. 81.

IEE analyzed the passivation effect of the Si:H films with different structures on c-Si wafer, as shown in Fig. 82. They found that H atoms in the form of Si–H bonds are more suitable to passivate the crystalline Si surface than those in the form of Si-H$_2$ bonds [234].

Chen et al. [235] also have done some work on the PECVD passivation of c-Si for HIT cells. They found that HF/O$_3$ cleaning improves the lifetime of the passivated wafers, as shown in Fig. 83.

Comment:	
Date/Time:	7/23/2011 09:35
Operator:	
Sample:	
Raster:	500 um
Size:	3 inch
Scanradius:	25 mm, 25 mm

u-PCD

Lifetime:	
Average:	**883.04 us**
Median:	891.25 us
Deviation:	37.027 %
Minimum:	85.149 us
Maximum:	1596.3 us

Time Range:	10 ms
Time Cursor:	Auto
Sensitivity:	200 mV
Averaging:	4
MW Freq.:	10.117 GHz
Laser Power:	120 E11
Pulse Width:	200 ns
Excited Area:	1 mm2
Laser Wavel.:	904 nm
Head Height:	0.65 mm
Duration:	00:10:24

400 us   1300 us

Fig. 80 Lifetime mapping of a c-Si wafer bifacially passivated by a-Si:H as obtained by the Nanchang University [231]

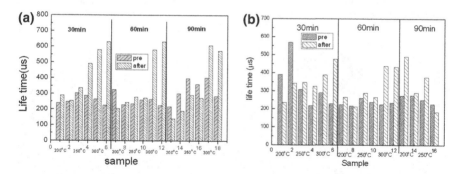

Fig. 81 a Lifetime of wafers passivated by a-Si:H before and after post-annealing, **b** lifetime of wafers passivated by a-SiO$_x$:H before and after post-annealing [232, 233]

10.2.3 Doped Layers: Emitter and BSF

Most of the doped silicon layers are made with PECVD and HWCVD. For an amorphous silicon layer, it is hard to get high conductivity with boron doping. With HWCVD, it is easier to get high doping efficiencies than with PECVD. In the case of the microcrystalline material, a high doping level can be obtained by high dilution in hydrogen. We only show a few examples of the production of these layers.

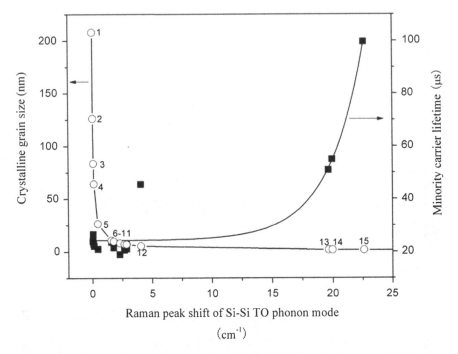

Fig. 82 The passivation effect of silicon thin films with different structures on c-Si surface, characterized by the minority carrier lifetime and the crystalline grain size in the silicon thin films as a function of the corresponding Raman peak shift of the Si–Si TO phonon mode between the silicon thin film and the c-Si substrate [234]

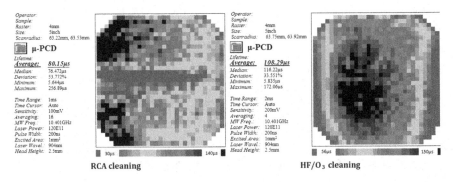

Fig. 83 Lifetime of wafers, cleaned with different recipes [244]

An optimum mc-Si:H p-layer with PECVD with a hydrogen dilution ratio of 99.65 %, an RF-power density of 0.08 W/cm^2, a gas-phase doping ratio of 0.125 %, and a p-layer thickness of 15 nm is obtained by the Graduate University of the Chinese Academy of Sciences [236]. The achieved optimized efficiency is 13.4 % and no texturization has been applied. An efficiency of 17.27 % has been

Fig. 84 Effect of hydrogen
dilution ratio (S_H) on the
properties of the solar cells.
[237]

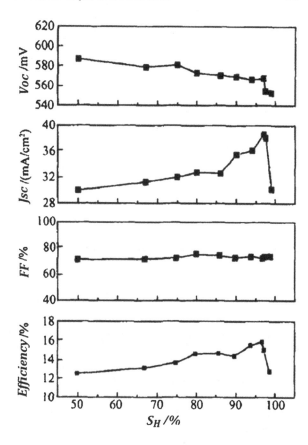

Fig. 84 Effect of hydrogen dilution ratio (S_H) on the properties of the solar cells. [237]

obtained by the same group with the HWCVD method as shown in Fig. 84 [237].
An n-type nanocrystalline Si:H film is used as emitter of the n-nc-Si:H/p-c-Si
heterojunction solar cells.

Minhua Wang and Bingyan Ren have done a series of investigations about p-
type a-Si:H deposited by RF-PECVD. They have produced (p)a-Si:H/(n)c-Si cells
with an efficiency of 9.8 % for optimized parameters [238].

The Nanchang University used a low deposition pressure for the p-type mc-
Si:H of only about 15 Pa, in order to get a low deposition rate required for high-
quality films. They used the films for the (p)mc-Si:H/(n) c-Si cells. The external
quantum efficiency measurement for one of the cells is shown in Fig. 85.

10.2.4 TCO and Metal Contacts

TCO and metal contact deposition are both very important topics in the semiconductor field. Many groups [239–242] have done work about them. Here, only
some exemplary works are shown.

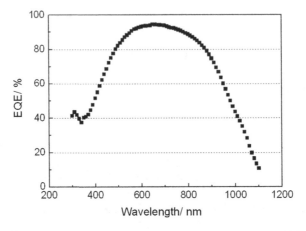

Fig. 85 The external quantum efficiency (EQE) measurement for a (p)mc-Si:H/(n)c-Si cell of the Nanchang University [226]

Cui et al. [242] in Shanghai Jiaotong University used the bias magnetron RF sputtering technique to produce ITO layers for heterojunction solar cells at a substrate temperature of 180 °C and a low substrate–target distance. They suggest that a bias voltage of −75 V is good for heterojunction solar cells.

Liu [243] at Hebei University of Technology studied the effect of post-annealing on the conductivity and transmittance of the ITO layers. Their best ITO layer shows a resistivity of 2.0×10^{-4} Ω cm and a transmittance larger than 90 %. They also made some solar cells to improve the efficiency, as shown in Fig. 86.

Some AZO and AZO/Ag/AZO multilayer structures for HIT solar cells, as shown in Figs. 87 and 88, have been tried by Chen et al. [230, 244] at Jiangnan University. In Ref. [244], it is stated that an Ar pressure of 4 mTorr and a power of

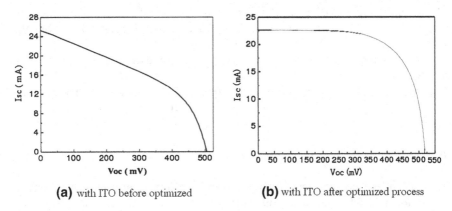

(a) with ITO before optimized   **(b)** with ITO after optimized process

Fig. 86 I–V curves of HIT solar cell before (**a**) and after (**b**) an optimized process, made by Liu [243]

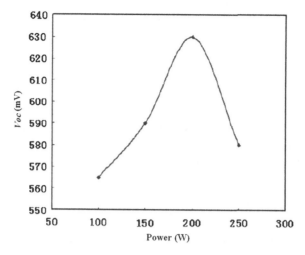

Fig. 87 Dependence of the V_{oc} on the sputtering power of the ZnO thin film [244]

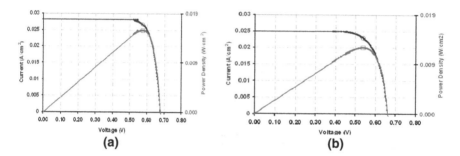

Fig. 88 I–V curves of the HIT cells tested by Suns-Voc, published in [230] **a** with AZO/Ag/AZO as front TCO layer; **b** with AZO as front TCO layer

200 W are best for the HIT cells. For these conditions, carrier collection efficiency is improved without compromising on the light absorption. In Ref. [230], it is stated that the AZO/Ag/AZO multilayer structure has an advantage as compared to the AZO single layer as front TCO layer for the HIT structure.

10.2.5 Cell Production

To the best of the author's knowledge, the highest confirmed efficiency of the a-Si:H/c-Si structure solar cells produced by Chinese researchers has been obtained by Wang [245]. The efficiency of a p-type c-Si wafer-based solar cell has been reported in this case to be 18.2 %. According to unconfirmed reports, the Shanghai Institute of Microsystem and Information Technology has recently achieved an efficiency of >19 % [246]. In the following, the cell fabrication situation will be presented.

n-type c-Si substrate

Wang et al. [230] used an AZO/Ag/AZO multilayer structure as the front TCO layer for the HIT cell. Their processes and parameters for the cells are shown in Fig. 89. Their best results are $V_{oc} = 677$ mV; $J_{sc} = 28.1$ mA/cm^2; and $\eta = 15.31$ %.

In 2006, GUCAS reported their result of (p)mc-Si:H/(n)c-Si cells [236]. In 2011, the same group found that the instability of the plasma in PECVD chamber in the initial stage would damage the surface of the c-Si. Using a shutter to shield the substrate for 100 s from the starting discharge can prevent the influence of the instable plasma process on the Si surface and also on the interface between a-Si:H and c-Si. It is also said that the effect of a hydrogen pre-treatment on the interfacial passivation is controlled by the extent of hydrogen plasma bombardment. The optimum time for hydrogen pre-treatment is about 60 s [247].

(a)

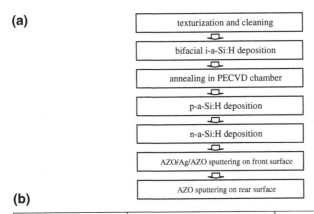

(b)

		Gas flow rate (sccm)						Power (W)	Pressure (Torr)	Depositing time (s)
		SiH$_4$	H$_2$	TMB	PH$_3$	Ar	O$_2$			
i-layer		3	30	—	—	—	—	12	1	91
p-layer		9	50	18	—	—	—	20	2	58
n-layer		6	40	—	20			25	2	105
AZO/Ag/AZO	Front AZO	—	—	—	—	18	0.15	160	0.004	300
	Ag	—	—	—	—	10	—	80	0.0035	57
	Rear AZO	—	—	—	—	18	—	160	0.004	300
AZO		—	—	—	—	18		—	0.003	720
Ag		—	—	—	—	10		—	0.003	1200

Fig. 89 **a** The processes sequence for HIT deposition, **b** the parameters for each layer, done by Wang [230]

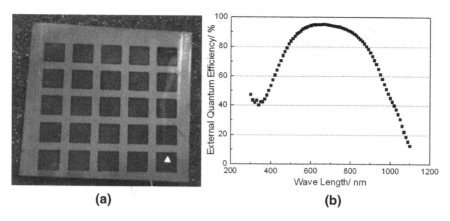

(a)   **(b)**

Fig. 90 **a** Photograph of the cell, **b** the external quantum efficiency of the cell made at Nanchang University [231]

The Nanchang University group has made a heterojunction solar cell with the structure: Al/ITO/(p)mc-Si:H/(i)a-Si:H/(n)c-Si/ITO/Al as shown in Fig. 90 in the year 2011 [231].

p-type c-Si substrate

In 2006, GUCAS reported their result for an (n)nc-Si:H/(p)c-Si heterojunction solar cell deposited by HWCVD [248]. The structure of the cell was ITO/(n)nc-Si:H/(i)a-Si:H/(p)c-Si/Al. The resulting solar cell parameters were $\eta = 10.2$ %, $J_{sc} = 29.5$ mA/cm^2, $V_{oc} = 483$ mV, and $FF = 70$ %.

In 2007, Liu [243] made a HIT cell with the structure Ag/ITO/(n)a-Si:H/(i)a-Si:H/(p)c-Si/Al. The efficiency was 13.38 % and the fill factor was 77 %.

In the year 2007, GUCAS reported an improvement for a (n)nc-Si:H/(p)c-Si structure [249]. They showed that hydrogen pre-treatment of the c-Si surface before film deposition is helpful to improve of the short-circuit current. In their experiments, using an optimized hydrogen pre-treatment time (30–60 s) by HWCVD, a gain of ~ 4 mA/cm^2 in short-circuit current was obtained compared to the cell without hydrogen treatment. They also showed the impact of the H$_2$ dilution for the intrinsic layer on the properties of the cells. The optimized H$_2$ dilution in their experiments was 97 %, resulting in a solar cell efficiency of 17.36 %. The total structure of this cell was Al/(n)nc-Si:H/(i)nc-Si:H/(p)c-Si/Al. Under this condition, epitaxial silicon growth was found at the interface of the (i)nc-Si:H/(p)c-Si, as shown in Fig. 91. So the authors concluded that the epitaxial layer is helpful in improving the solar cell [249], which is opposed to the results from the Ref. [250]. GUCAS has studied the details of the solar cells with C–V, C-f method, admittance spectroscopy, and dark I–V [249, 251]. They also found that low-temperature post-annealing of the solar cell improves V_{oc}, FF, and η [223].

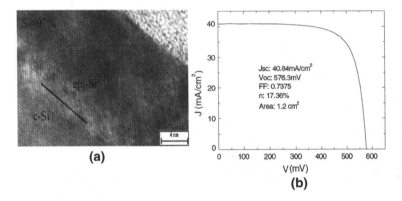

Fig. 91 a TEM image of the buffer layer prepared at a hydrogen dilution ratio $S_H = 97$ %; **b** *J*–*V* curve of an (n)epi-Si:H/(i)nc-Si:H/(p)c-Si solar cell [249] (*With kind permission from Springer Science and Business Media.*)

10.3 Theoretical Simulation

The theoretical analysis of the c-Si-based heterojunction solar cells began in the University of Science and Technology of China, to the knowledge of the authors. Lin and Che [252] published some articles for the analysis of the a-Si/c-Si heterojunction solar cells in 1997.

At the beginning, the researchers focused on the ideal prototype pn or pin structure cells, in which each layer has uniform properties and neglected the interface defect densities [252, 253]. Then, they cared about some details such as BSF [254], band offsets (ΔE_C, ΔE_V) [255], and density of the interface states [256].

The main theoretical analysis and conclusions from the Chinese researcher will be presented in the following in historical order.

Lin et al. [252, 253, 257] used a homemade program to do the analysis. The theory and solvers which they used were standard tools. Most of their analysis was done on n-type c-Si-based cells. They analyzed on the (p)a-Si:H/(n)c-Si structure [252, 253], (p)a-Si:H/(i)a-Si:H/(n)c-Si structure [252, 253, 257, 258], and (p)a-SiC/(n)c-Si structure [259, 260]. They took into account the band structures and electric field distribution under thermal equilibrium. They came to the following conclusions:

1. When an intrinsic layer is inserted between the (p)a-Si:H layer and the (n)c-Si substrate, the electric field in the amorphous silicon region will be enhanced, mostly to values larger than 10^4 V/cm. This will reduce the "dead area" where the photo-carriers cannot be pulled out by the electric field.
2. The light-induced degradation is small in the (p)a-Si:H/(i)a-Si:H/(n)c-Si structure as shown in Figure 92, although there are sensitive a-Si:H layers and

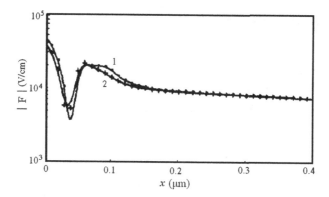

Fig. 92 Electric field profiles in the (p)a-Si:H/(i)a-Si:H/(n)c-Si structure before (*line 1*) and after (*line 2*) light soaking [252]

the light causes some charged defects in the space charge area. These charged defects will only slightly affect the drift of the photo-carriers in the (p)a-Si:H/(i)a-Si:H/(n)c-Si structure.

3. It is advantageous to use a (p)a-SiC:H film instead of a (p)a-Si:H film as emitter of the heterojunction solar cells. The emitter of the heterojunction should be thin enough.

   Zhao et al. [185, 254, 261–263] used AFORS-HET to analyze the parameters of each layer in (n)a-Si:H/(i)a-Si:H/(n)c-Si heterojunction solar cells, as shown in Fig. 93. They drew the following conclusions:

1. It was predicted that the polymorphous silicon capable of producing the optimal BSF effect is the microcrystalline silicon with a bandgap of 1.6 eV, a doping concentration of 10^{18} cm^{-3} and a thickness about 5 nm. Li et al. [264] of the Inner Mongolia Normal University came to a very similar conclusion. Such microcrystalline silicon BSF is easy to fabricate. It renders the efficiency of solar cells much higher than that using the conventional Al BSF with the same doping concentration. The thickness of the BSF has a negligible effect on the properties of the cells.

2. The work function of the TCO layer obviously affects the HIT performance, its value should be as low as possible for the front TCO/(n)a-Si:H/(i)a-Si:H/(p)c-Si structure. If the work function of the TCO layer is not appropriate, the built-in potentials of the TCO/a-Si:H contact and the a-Si:H/c-Si junction will have directions opposite to each other.

   It is stated that the n$^+$ layer as the BSF of the (n)c-Si-based heterojunction solar cell improves the ohmic contact to the metal. The spectral response of the solar cell in the long-wavelength region is improved by the BSF structure. With

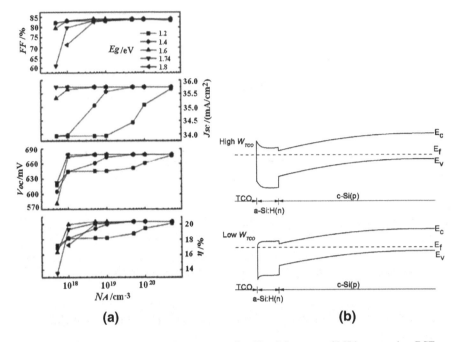

Fig. 93 a Effect of E_g and the doping concentration N_A of the p-type Si:H layer used as BSF on the properties of the HIT cells [254], **b** Energy band diagrams of the TCO/a-Si:H(n)/a-Si:H(i)/c-Si(p) structure with a low or a high W_{TCO} (work function). E_C is the edge of conduction band, E_f the Fermi level, and E_V is the edge of the valence band [262]

increasing doping density of the n^+ layer, the performance of the solar cells becomes better significantly. [265]

Zhong et al. [191, 255, 266] analyzed the physical mechanism of the S-shaped J–V characteristics (such as Fig. 94a) of (p)a-Si:H/(n)c-Si heterojunction solar cells at low temperatures by means of AMPS, some results are shown in Fig. 94. They assumed a low impurity concentration in the a-Si:H layer, a high valence band offset, and high interface defect densities. The results showed that the barrier at the amorphous/crystalline interface hindered the collection of the photogenerated holes. A high hole accumulation at the interface in combination with the interface defects and electrons caused a shift of the depletion region from the c-Si into the a-Si:H layer. This led to a decreased electric field and enhanced recombination inside the c-Si depletion region and in turn to a significant current loss.

An analytical expression for the diffusion capacitance C_D has been developed by Zhong. It is stated that C_D decreases with increasing interface defect density, D_{it}, because interface states act as recombination centers and decrease the excess carrier density in c-Si. C_D increases with decreasing D_{it} down to 10^{10} cm^{-2} eV^{-1}. As a result, the measurement is sensitive to D_{it} values down to 10^{10} cm^{-2}eV^{-1}. Accordingly, interface states can be characterized directly by the theoretical diffusion capacitance. The band offsets affect the surface recombination velocity and

Fig. 94 a Examples of the S-shaped I–V curve and the conventional *I-V* curve [245]; **b** Band structure of thermal equilibrium situation and short current under light-induced situation for different doping concentrations (N_A) of the emitter [266]; **c** Simulation results of the light I–V curve of the heterojunction solar cells with different N_A of the emitter [266]

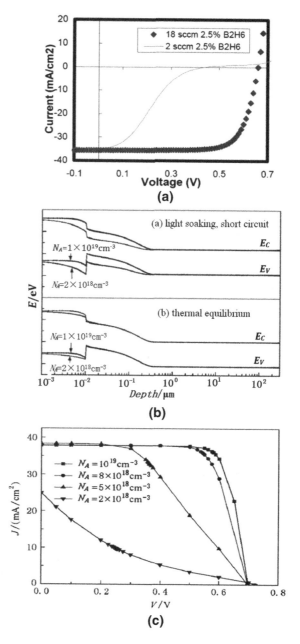

the carrier transport, and thus the I–V characteristics. For non-negligible interface states, their effects on the I–V characteristics of (p)a-Si:H/(n) c-Si heterojunctions under illumination are gradually eliminated with increasing (≤ 0.5 eV) valence band offsets, $\ddot{A}E_V$. As a result, appropriate band offsets can improve the efficiencies of a-Si:H/c-Si heterojunctions [255].

References

1. Grigorovici R, Croitoru N, Marina M, Nastase L (1968) Heterojunctions between amorphous Si and Si single crystals. Rev Roum Phys 13:317–325
2. Grigorovici R, Croitoru N, Teleman E, Marina M (1965) Rev Roum Phys 10:641–659
3. Jayadevaiah T, Busmundrud O (1972) Amorphous-crystalline silicon junctions. Electron Lett 8:75–77
4. Neitzert HC, Hirsch W (1995) In-situ measurements of changes of the excess charge carrier decay rate at the crystalline silicon surface during low temperature annealing and cooling. Phys Status Solidi A 151:371–377
5. Fuhs W, Niemann K, Stuke J (1974) Heterojunctions of amorphous silicon and silicon single crystals. Bull Am Phys Soc 19:345–350
6. Mott NF (1972) Conduction in non-crystalline materials. J Non-Cryst Solids 8–10:1–18
7. Dunn B, Mackenzie JD, Clifton JK, Masi JV (1975) Heterojunction formation using amorphous materials. Appl Phys Lett 26:85–86
8. Brodsky MH, Döhler GH, Steinhardt PJ (1975) On the Measurement of the conductivity density of states of evaporated amorphous silicon films. Phys Status Solidi B 72:761–770
9. Spear W, Le Comber P (1975) Substitutional doping of amorphous silicon. Solid State Commun 17:1193–1196
10. Matsuura H, Okuno T, Okushi H, Tanaka K (1984) Electrical properties of n-amorphous/p-crystalline silicon heterojunctions. J Appl Phys 55:1012–1019
11. Sasaki G, Fujita S, Sasaki A (1982) Gap-states measurement of chemically vapor-deposited amorphous silicon: high-frequency capacitance-voltage method. J Appl Phys 53:1013–1017
12. Klausmann E, Fahrner WR, Loeffler S, Neitzert HC (1993) Comparison of lifetime measurements from the Zerbst and the dispersion techniques. J Electrochem Soc 140:2323–2327
13. Mimura H, Hatanaka Y (1987) The use of amorphous-crystalline silicon heterojunctions for the application to an imaging device. J Appl Phys 61:2575–2580
14. Okuda K, Okamoto H, Hamakawa Y (1983) Amorphous Si/polycrystalline Si stacked solar cell having more than 12% conversion efficiency. Jpn J Appl Phys 22:L605–L607
15. Nozagi H, Hatayama T, Ito H, Ide K, Nakagawa M (1984) Amorphous Si/ribbon Si tandem type solar cell. In: Proceedings of the PVSEC-1, Kobe
16. Kunst M, Beck G (1988) The study of charge carrier kinetics in semiconductors by microwave conductivity measurements II. J Appl Phys 63:1093–1098
17. Kunst M, Werner A (1985) Charge carrier dynamics in a-Si:H. Solid State Commun 54:119–121
18. Neitzert HC, Kunst M (1992) Microwave detected transient photoconductivity measurements during plasma deposition of intrinsic hydrogenated amorphous silicon. Appl Phys A: Mater Sci Process 55:378–386

W. R. Fahrner, *Amorphous Silicon / Crystalline Silicon Heterojunction Solar Cells*,
SpringerBriefs in Applied Sciences and Technology, DOI: 10.1007/978-3-642-37039-7,
© Chemical Industry Press, Beijing and Springer-Verlag Berlin Heidelberg 2013

19. Neitzert HC, Kunst M, Hirsch W, Swiatkowski C (1990) In-situ investigation of optoelectronic properties of cristalline silicon/amorphous silicon heterojunctions. In: Proceedings of the PVSEC-5, Kyoto, pp 825–827

20. Neitzert HC, Hirsch W, Kunst M (1993) Structural changes of a-Si:H films on crystalline silicon substrates during deposition. Phys Rev B 47:4080–4083

21. Neitzert HC, Layadi N, Roca i Cabarrocas P, Vanderhaghen R, Kunst M (1995) Laser pulse induced microwave conductivity and spectroscopic elipsometry characterization of helium and hydrogen plasma damage of the crystalline silicon surface. Mater Sci Forum 173–174:209–214

22. Neitzert HC, Hirsch W, Kunst M (1993) Transfer of excess charge carriers in an a-Si:H/ crystalline-silicon heterojunction measured during the growth of the amorphous silicon layer. Phys Rev B 48:4481–4486

23. Neitzert H (1993) Characterization of electronic transport in amorphous silicon/crystalline silicon photodiodes. In: SPIE proceedings, vol. 1985. Trieste, pp 530–540

24. Takahama T, Tagutchi M, Kuroda S, Matsuyama M, Tanaka M, Tsuda S, Nakano S, Kuwano Y (1992) High efficiency single- and poly-crystalline silcon solar cells using ACJ-HIT structure. In: Proceedings of the 11th European photovoltaic solar energy conference, Montreux, pp 1057–1060

25. Tanaka M, Taguchi M, Matsuyama T, Sawada T, Tsuda S, Nakano S, Hanafusa H, Kuwano Y (1992) Development of new a-Si/c-Si heterojunction solar cells: ACJ-HIT (artificially constructed junction-heterojunction with intrinsic thin-layer). Jpn J Appl Phys 31: 3518–3522

26. Sanyo (2009) SANYO develops HIT solar cells with world's highest energy conversion efficiency of 23.0%. http://us.sanyo.com/News/SANYO-Develops-HIT-Solar-Cells-with-World-s-Highest-Energy-Conversion-Efficiency-of-23-0-. Accessed 21 May 2009

27. http://www.solarwirtschaft.de

28. Piria R, Urbschat C, Mueller S www.bee-ev.de/uploads/BEE-Studie%20Weltmarkt potenzial%202020.pdf

29. Sanyo Electric Co. Ltd (2009) SANYO to build new HIT solar cell production facilities at Nishikinohama factory. http://solarcellsinfo.com/blog/archives/2133. Accessed 16 Feb 2009

30. Osborne M (2009) Sanyo targets 600MW HIT solar cell production with new plant. http://www.pvtech.org/news/_a/sanyo_targets_600MW_hit_solar_cell_production_with_new_plant/. Accessed 17 Feb 2009

31. Technische Daten Sanyo HIP-225HDE1 http://www.solarshop.net/solarkomponenten/solarmodule/sanyo_hip-225hde1_m_666.html

32. Terakawa A, Asaumi HT, Kobayashi S, Tsunomura Y, Yagiura T, Taguchi M, Yoshimine Y, Sakata H, Maruyama E, Tanaka M (2005) High efficiency HIT^{TM} solar cells and effects of open circuit voltage on temperature coefficients. In: Proceedings of the 15th PVSEC, Shanghai, p 661

33. Taguchi M, Maruyama E, Tanaka M (2008) Temperature dependence of amorphous/crystalline silicon heterojunction solar cells. Jpn J Appl Phys 47:814–818

34. Hake MG (2008) HIT technology. http://www.sanyosolar.eu/fileadmin/EDITORS/PHOTOVOLTAICS/PRESENTATIONS/080602_HIT_presentation.pdf last visited 2009-05025. Accessed in 2008

35. Runyan WR (1965) Silicon semiconductor technology. McGraw-Hill, New York

36. Goetzberger A, Knobloch J, Voss B (1998) Crystalline silicon solar cells. Wiley, New York

37. Hull R (1999) Properties of crystalline silicon. Inspec, IEE, London

38. Kamins T (1988) Polycrystalline silicon for integrated circuits applications. Kluwer Academic Publishers, Norwell

39. Springer J, Poruba A, Fejfar A, Vanecek M, Feitknecht L, Wyrsch N, Meier J, Shah A (2001) Nanostructured thin film silicon solar cells: optical model. In: Proceedings of the 16th photovoltaic solar energy conference Glasgow: 2000, pp 434–437

40. Hapke P (1995) VHF-plasma deposition of microcrystalline silicon (μc-Si:H): influence of the plasma excitation frequency on the structural and electrical properties. PhD thesis, Aachen, Germany

41. Roth A (1993) Preparation of amorphous and microcrystalline silicon thin layers by means of the hydrogen abstraction method. Frankfurt/Main, Germany

42. Street RA (1991) Hydrogenated amorphous silicon. Cambridge University Press, Cambridge

43. Janssen R, Janotta A, Dimova-Malinovska D, Stutzmann M (1999) Optical and electrical properties of doped amorphous silicon suboxides. Phys Rev B 60:13561–13572

44. Tauc J (1974) Amorphous and liquid semiconductors. Plenum Press, New York

45. Davis EA, Mott NF (1970) Conduction in non-crystalline systems V: conductivity, optical absorption and photoconductivity in amorphous semiconductors. Philos Mag 22:903–922

46. Joannopoulos JD, Lucovsky G, Knights JC, Spear WE, LeComber PG, Thompson MJ, Kaplan D, Carlson DE, Madan A (1984) The physics of hydrogenated amorphous silicon. Springer, Berlin

47. Kanicki J (1992) Amorphous and microcrystalline semiconductor devices—materials and device physics. Artech House, Boston

48. Searle T (1998) Amorphous silicon and its alloys. Inspec, IEE, London

49. Heck S (2002) Investigation of light-induced defects in hydrogenated amorphous silicon by low temperature annealing and pulsed degradation. PhD thesis, Marburg, Germany

50. Evans CJ, Paul E, Dornfeld D, Lucca DA, Byrne G, Tricard M, Klocke F, Dambon O, Mullany BA (2003) Material removal mechanisms in lapping and polishing. CIRP Ann-Manuf Technol 52:611–633

51. Marinescu ID, Uhlmann E, Doi T (2006) Handbook of lapping and polishing. CRC Press, Boca Raton

52. Pei ZJ, Fisher GR, Liu J (2008) Grinding of silicon wafers: A review from historical perspectives. Int J Mach Tools Manuf 48:1297–1307

53. Zhao J, Green MA (1991) Optimized antireflection coatings for high-efficiency silicon solarcells. IEEE Trans Electron Devices 38:1925–1934

54. Pla J, Tamasi M, Rizzoli R, Losurdo M, Centurioni E, Summonte C, Rubinelli F (2003) Optimization of ITO layers for applications in a-Si/c-Si heterojunction solar cells. Thin Solid Films 425:185–192

55. Scherff ML, Stiebig H, Schwertheim S, Fahrner WR (2006) Double layer antireflexion coating of heterojunction solar cells. In: Proceedings of the 21st European photovoltaic solar energy conference, Dresden, Germany, pp 1392–1395

56. Granqvist CG (2007) Transparent conductors as solar energy materials: a panoramic review. Sol Energy Mater Sol Cells 91:1529–1598

57. Green MA (1987) High efficiency silicon solar cells. Trans Tech Publications, Aedermannsdorf

58. Gangopadhya U, Dutta SK, Saha H (2009) Texturization and light trapping in silicon solar cells. Nova Science Publisher Inc., Bangalore

59. Campbell P, Wenham SR, Green MA (1988) Light trapping and reflection control with tilted pyramids and grooves. In: Proceedings of the 20th IEEE photovoltaic specialists conference, Las Vegas, USA, pp 713–716

60. Sontag D, Hahn G, Fath P, Bucher E, Krühler W (2003) Texturing techniques and resulting solar cell parameters on Tri-silicon material. In: Proceedings of the 3rd world conference on photovoltaic energy conversion, Osaka, Japan, pp 1304–1307

61. Gerhards C, Fischer A, Fath P, Bucher E (2000) Mechanical microtexturization for multicrystalline Si solar cells. In: Proceedings of the 16th European photovoltaic solar energy conference, Glasgow, UK, pp 1390–1393

62. Inomata Y, Fukui K, Shirasawa K (1997) Surface texturing of large area multicrystalline silicon solar cells using reactive ion etching method. Sol Energy Mater Sol Cells 48:237–242

63. Winderbaum S, Reinhold O, Yun F (1997) Reactive ion etching (RIE) as a method for texturing polycrystalline silicon solar cells. Sol Energy Mater Sol Cells 46:239–248

64. Dekkers HFW, Duerinckx F, Szlufcik J, Nijs J (2000) Silicon surface texturing by reactive ion etching. Opto-Electron Rev 8:311–316

65. Zaidi SH, Ruby DS, Gee JM, Inc G, Albuquerque NM (2001) Characterization of random reactive ion etched-textured silicon solar cells. IEEE Trans Electron Devices 48:1200–1206

66. Ruby DS, Zaidi S, Narayanan S, Yamanaka S, Balanga R (2005) RIE-texturing of industrial multicrystalline silicon solar cells. J Sol Energy Eng 127:146–149

67. Lee DB (1969) Anisotropic etching of silicon. J Appl Phys 40:4569–4574

68. Macdonald DH, Cuevas A, Kerr MJ, Samundsett C, Ruby D, Winderbaum S, Leo A (2004) Texturing industrial multicrystalline silicon solar cells. Sol Energy 76:277–283

69. Green MA (1995) Silicon solar cells: advanced principles and practice. Centre for Photovoltaic Devices and Systems, Sydney

70. Brendel R, Auer R, Bail M, Feldrapp K, Horbelt R, Kintzel W, Kuchler G, Mueller G, Scholten D, Artmann H (2001) Monokristalline Waffelzellen aus Silizium nach dem perforierten Silizium (PSI)-Prozess. Final report for BMWi from ZAE Bayern

71. You JS, Kim D, Huh JY, Park HJ, Pak JJ, Kang CS (2001) Experiments on anisotropic etching of Si in TMAH. Sol Energy Mater Sol Cells 66:37–44

72. Angermann H, Roeseler A, Rebien M, Henrion W (2004) Wet-chemical preparation and spectroscopic characterization of Si interfaces. Appl Surf Sci 235:322–329

73. Jensen N (2002) Heterostruktursolarzellen aus amorphem und kristallinem Silicium. PhD thesis, Stuttgart University, p 129

74. Hattori T (1998) Ultraclean surface processing of silicon wafers. Springer, Berlin

75. Liebermann MA, Lichtenberg AJ (1994) Principles of plasma discharges and materials processing. Wiley, New York

76. Fujiwara H, Kondo M (2006) Interface structure in a-Si:H/c-Si heterojunction solar cells characterized by optical diagnoses technique. In: Proceedings of the 4th IEEE world conference on photovoltaic energy conversion (WCPEC-4), Waikoloa, Hawaii, USA, pp 1443–1448

77. Luque A, Hegedus S (2003) Handbook of photovoltaic science and engineering. Wiley, New York

78. Zhang D, Shahbazi S, Zhang W, Moyer J, Guo T (2008) A screen printable front side silver conductor paste achieving high aspect ratio finger lines for solar cell application. In: Proceedings of the 23rd European photovoltaic solar energy conference, Valencia, Spain, pp 1458–1460

79. Wünsch F, Klein D, Podlasly A, Ostmann A, Schmidt M, Kunst M (2009) Low-temperature contacts through SixNy-antireflection coatings for inverted a-Si:H/c-Si hetero-contact solar cells. Sol Energy Mater Sol Cells 93:1024–1028

80. Tucci M, Salurso E, Roca F, Palma F (2002) Dry cleaning process of crystalline silicon surface in a-Si:H/c-Si heterojunction for photovoltaic applications. Thin Solid Films 403–404:307–311

81. Borchert D, Brammer T, Voigt O, Stiebig H, Gronbach A, Rinio M, Kenanoglu A, Willeke G, Windgassen H, Nositschka WA, Kurz H (2004) Large area (n) a-Si:H/(p) c-Si heterojunction solar cells with low temperature screen printed contacts. In: Proceedings of the 19th European photovoltaic solar energy conference, Paris, France, pp 584–587

82. Jagannathan B, Anderson W, Coleman J (1997) Amorphous silicon/p-type crystalline silicon heterojunction solar cells. Sol Energy Mater Sol Cells 46:289–310

83. Hussein MR (2000) Herstellung und Charakterisierung von großflächigen Heterosolarzellen. Master thesis, University of Hagen

84. Schröder B, Weber U, Ledermann A, Seitz H, Kupich M, Mukherjee C (2001) Progress in thin-film-silicon-based solar cells prepared by thermo-catalytic CVD. In: Proceedings of the 17th European photovoltaic conference, Muenchen, p 2850

85. Tsuda S, Takahama T, Hishikawa Y, Nishiwaki H, Wakisaka K, Nakano S (1993) a-Si technologies for high efficiency solar cells. J Non-Cryst Solids 164–166:679–684

86. Sawada T, Terada N, Tsuge S, Baba T, Takahama T, Wakisaka K, Tsuda S, Nakano S (1994) High efficiency a-Si/c-Si heterojunction solar cell. In: Proceedings of the 24th IEEE photovoltaic specialist conference, Waikaloa, Hawaii, pp 1219–1226

87. Schmidt M, Angermann H, Conrad E, Korte L, Laades A, von Maydell K, Schubert C, Stangl R (2006) Physical and technological aspects of a-Si:H/c-Si hetero-junction solar cells 2006. In: Proceedings of the 4th IEEE world conference on photovoltaic energy conversion, Waikaloa, Hawaii, pp 1433–1438

88. Conrad E, Maydell KV, Angermann H, Schubert C, Schmidt M (2006) Optimization of interface properties in a-Si:H/c-Si heterojunction solar cells. In: Proceedings of the 2006 IEEE 4th world conference on photovoltaic energy conversion, Waikaloa, USA, pp 1263–1266

89. Cleef MWMV, Rubinelli FA, Daey Ouwens JD, Schropp R (1995) Amorphous-crystalline heterojunction silicon solar cells with an a-SiC:H window layer. In: Proceedings of the 13th European photovoltaic solar energy conference, Nice, France, pp 1303–1306

90. Martin I, Vetter M, Garin M, Orpella A, Voz C, Puigdollers J, Alcubilla R (2005) Crystalline silicon surface passivation with amorphous SiC_x:H films deposited by plasma-enhanced chemical-vapor deposition. J Appl Phys 98:114912–10

91. Wünsch F, Citarella G, Abdallah O, Kunst M (2006) An inverted a-Si:H/c-Si hetero-junction for solar energy conversion. J Non-Cryst Solids 352:1962–1966

92. Hezel R, Jäger K (1983) Properties of inversion layers for MIS/IL solar cells studied on low-temperature-processed MNOS transistors. Solid-State Electron 26:993–997

93. Ulyashin AG, Job R, Scherff M, Gao M, Fahrner WR, Lyebyedyev D, Roos N, Scheer H-C (2002) The influence of the amorphous silicon deposition temperature on the efficiency of the ITO/A-Si:H/C-Si heterojunction (HJ) solar cells and properties of interfaces. Thin Solid Films 403:359–362

94. Stangl R, Froizheim A, Schmidt M, Fuhs W (2003) Design criteria for amorphous/cristalline silicon heterojunction solar cells—a simulation study. In: Proceedings of the WCPEC-3, Osaka, pp 1005–1008

95. Stangl R, Haschke J, Bivour M, Korte L, Schmidt M, Lips K, Rech B (2009) Planar rear emitter back contact silicon heterojunction solar cells. Sol Energy Mater Sol Cells 93:1900–1903

96. Tucci M, Serenelli L, Salza E, De Iuliis S, Geerligs L, Caputo D, Ceccarelli M, de Cesare G (2008) Back contacted a-Si:H/c-Si heterostructure solar cells. J Non-Cryst Solids 354:2386–2391

97. Lu M, Bowden S, Das U, Birkmire R (2007) a-Si/c-Si heterojunction for interdigitated back contact solar cell. In: Proceedings of the 22nd European photovoltaic solar energy conference, Milano, Italia, pp 20–23

98. Johnson JN, Manivannan V (2008) Photovoltaic device which includes all-back-contact configuration and related fabrication process. US patent application no. US2008/0000522

99. Terakawa A, Asaumi T (2008) Photovoltaic Cell and method of fabricating the same. U.S. patent no. US 7,199,395 B2

100. Herrmann H, Herzer H, Sirtl E (1975) Modern silicon technology. Festkörperprobleme, vol 15. Springer, Berlin, pp 279–316

101. Rubinelli F (1987) Amorphous-crystalline silicon anisotype heterojunctions: built-in potential, its distribution and depletion widths. Solid-State Electron 30:345–351

102. Cuniot M, Marfaing Y (1988) Energy band diagram of the a-Si:H/c-Si interface as determined by internal photoemission. Philos Mag Part B 57:291–300

103. Dingle R (1975) Optical and electronic properties of thin AlxGa1-x As/GaAs heterostructures. Crit Rev Solid State Mater Sci 5:585–590

104. Miller RC, Kleinman DA, Gossard AC (1984) Energy-gap discontinuities and effective masses for $GaAs-Al_xGa_{1-x}As$ quantum wells. Phys Rev B 29:7085–7087

105. Kroemer H (1993) Semiconductor heterojunctions at the conference on the physics and chemistry of semiconductor interfaces: a device physicist's perspective. In: Proceedings of the 20th annual conference on the physics and chemistry of semiconductors interfaces, Williamsburg, USA, pp 1354–1361

106. Sawada T, Terada N, Tsuge S, Baba T, Takahama T, Wakisaka K, Tsuda S, Nakano S (1994) High-efficiency a-Si/c-Si heterojunction solar cell. In: Proceedings of 1994 IEEE 1st world conference on photovoltaic energy conversion, Waikoloa, USA, pp 1219–1226

107. Jensen N, Hausner RM, Bergmann RB, Werner JH, Rau U (2002) Optimization and characterization of amorphous/crystalline silicon heterojunction solar cells. Prog Photovoltaics: Res Appl 10:1–13

108. Fahrner WR, Mueller T, Scherff M, Knoener D, Neitzert HC (2006) Interface states of heterojunction solar cells. In: Proceedings of 4th world conference photovoltaic energy conversion, pp 1822–1825

109. Nicollian EH, Goetzberger A (1966) The Si-SiO2 interface—electrical properties as determined by the metal-insulator-silicon conductance technique. Bell Syst Tech J XLV I:1055

110. Sharma DK, Narasimhan KL (1991) Analysis of high-frequency capacitance of amorphous silicon-crystalline silicon heterojunctions. Philos Mag B 63:543–550

111. Gudovskikh AS, Kleider JP (2007) Capacitance spectroscopy of amorphous/crystalline silicon heterojunction solar cells at forward bias and under illumination. Appl Phys Lett 90:034104

112. Gudovskikh AS, Kleider JP, Damon-Lacoste J, Roca i Cabarrocas P, Veschetti V, Muller JC, Ribeyron PJ, Rolland E (2006) Interface properties of a-Si:H/c-Si heterojunction solar cells from admittance spectroscopy. Thin Solid Films 511–512:385–389

113. Street RA (1991) Hydrogenated amorphous silicon. Cambridge University Press, Cambridge

114. Roesch M (2003) Experimente und numerische Modellierung zum Ladungstraegertransport in a-Si:H/c-Si Heterodioden. Ph. D. thesis, University of Oldenburg

115. Taguchi M, Terakawa A, Maruyama E, Tanaka M (2005) Obtaining a higher VOC in HIT cells. Prog Photovoltaics: Res Appl 13:481–488

116. Kerr MJ, Cuevas A (2002) Very low bulk and surface recombination in oxidized silicon wafers. Semicond Sci Technol 17:35–38

117. Kerr MJ, Cuevas A (2002) Recombination at the interface between silicon and stoichiometric plasma silicon nitride. Semicond Sci Technol 17:166–172

118. Schmidt J (1998) Untersuchungen zur Ladungstraegerrekombination an den Oberflaechen und im Volumen von kristallinen Silicium-Solarzellen. PhD thesis, Hannover University

119. Taguchi M, Kawamoto K, Tsuge S, Baba T, Sakata H, Morizane M, Uchihashi K, Nakamura N, Kiyama S, Oota O (2000) HIT cells—high efficiency crystalline Si cells with novel structure. Prog Photovoltaics 8:503–513

120. Martin I, Vetter M, Qorpella J, Puigdollers J, Cuevas A, Alcubilla R (2001) Surface passivation of p-type crystalline Si by plasma enhanced chemical vapor deposited amorphous SiCx:H films. Appl Phys Lett 79:2199–2201

121. Garin M, Rau U, Brendle W, Martin I, Alcubilla R (2005) Characterization of a-Si:H/c-Si interfaces by effective-lifetime measurements. J Appl Phys 98:093711

122. Vetter M, Touati Y, Martin I, Ferre R, Alcubilla R, Torres I, Alonso J, Vazquez M (2005) Characterization of industrial p-type CZ silicon wafers passivated with a-SiCx:H films. In: Proceedings of the 2005 IEEE Spanish conference on electronic devices (CDE05), Tarragona, Spain, pp 247–250

123. Olibet S, Vallat-Sauvain E, Ballif C (2007) Model for a-Si:H/c-Si interface recombination based on the amphoteric nature of silicon dangling bonds. Phys Rev B 76:035326

124. Laades A, Klieforth K, Korte L, Brendel W, Stangl R, Schmidt M, Fuhs W (2004) Surface passivation of crystalline silicon wafers by hydrogenated amorphous silicon probed by time resolved surface photovoltage and photoluminescence spectroscopy. In: Proceedings of the 19th European photovoltaic solar energy conference, Paris, France, pp 1170–1173

125. Wolf SD, Beaucarne G (2006) Surface passivation properties of boron-doped plasma-enhanced chemical vapor deposited hydrogenated amorphous silicon films on p-type crystalline Si substrates. Appl Phys Lett 88:022104

126. Dauwe S, Schmidt J, Hezel R (2002) Very low surface recombination velocities on p- and n-type Silicon wafers passivated with hydroginated amorphous silicon films. In: Proceedings of the 29th IEEE photovoltaic specialists conference, New Orleans, pp 1246–1249

127. Wang T, Iwaniczko E, Page M, Levi D, Yan Y, Yelundur V, Branz H, Rohatgi A, Wang Q (2005) Effective interfaces in silicon heterojunction solar cells. In: Proceedings of the 31st IEEE photovoltaic specialists conference, Orlando, USA, pp 955–958

128. Maruyama E, Terakawa A, Taguchi M, Yoshimine Y, Ide D, Baba T, Shima M, Sakata H, Tanaka M (2006) Sanyo's challenges to the development of high-efficiency HIT solar cells and the expansion of HIT business. In: Proceedings of the 4th IEEE world conference on photovoltaic energy conversion, Waikoloa, Hawaii, USA, pp 1455–1460

129. Fujiwara H, Kondo M (2007) Effects of a-Si:H layer thicknesses on the performance of a-Si:H/c-Si heterojunction solar cells. J Appl Phys 101:054516

130. Mueller T, Duengen W, Job R, Scherff M, Fahrner WR (2007) Crystalline silicon surface passivation by PECV deposited hydrogenated amorphous silicon oxide films (a-SiOx:H). In: Proceedings of the materials research society: symposium A, pp A05–02

131. Mueller T, Schwertheim S, Scherff M, Fahrner WR (2008) High quality passivation for heterojunction solar cells by hydrogenated amorphous silicon suboxide films. Appl Phys Lett 92:033504

132. Fujiwara H, Kaneko T, Kondo M (2007) Application of hydrogenated amorphous silicon oxide oxide layers to c-Si heterojunction solar cells. Appl Phys Lett 91:133508

133. Korevaar BA, Fronheiser J, Zhang X, Fedor LM, Tolliver TR (2008) Influence of annealing on performance for heterojunction a-Si/c-Si devices. In: Proceedings of the 23rd European photovoltaic solar energy conference, Valencia, Spain, pp 1859–1962

134. Kanno H, Ide D, Tsunomura Y, Taira S, Baba T, Yoshimine Y, Taguchi M, Kinoshita T, Sakata H, Maruyama E (2008) Over 22 % efficient HIT solar cell. In: Proceedings of the 23rd European photovoltaic solar energy conference and exhibition, Valencia, Spain, pp 1136–1139

135. Street RA (2000) Technology and applications of amorphous silicon. Springer, New York

136. Hamakawa Y (1979) Present status of solar photovoltaic R&D projects in Japan. Surf Sci 86:444–461

137. Nitta Y, Okamoto H, Hamakawa Y (1980) Amorphous Si: heteroface photovoltaic cells based upon p-i-n junction structure. Jpn J Appl Phys 19:143–148

138. Okamoto H, Yamaguchi T, Nitta Y, Hamakawa Y (1980) Effect of DC electric field on the basic properties of RF plasma deposited a-Si. J Non-Cryst Solids 35–36:201–206

139. Okamoto H, Nitta Y, Yamaguchi T, Hamakawa Y (1980) Device physics and design of a-Si ITO/p-i-n heteroface solar cells. Sol Cell Mater 2:313–325

140. Mueller T, Duengen W, Ma Y, Job R, Scherff M, Fahrner WR (2007) Investigation of the emitter band gap widening of heterojunction solar cells by use of hydrogenated amorphous carbon silicon alloys. J Appl Phys 102:074505

141. Mueller T (2009) Heterojunction solar cells (a-Si/c-Si): investigations on PECV deposited hydrogenated silicon alloys for use as high-quality surface passivation and emitter/BSF. PhD thesis, Hagen

142. Kunst M, Werner A (1985) Comparative study of time-resolved conductivity measurements in hydrogenated amorphous silicon. J Appl Phys 58:2236–2241

143. Schieck R, Kunst M (1997) Frequency modulated microwave photoconductivity measurements for characterization of silicon wafers. Solid-State Electron 41:1755–1760

144. Swiatkowski C, Sanders A, Buhre K, Kunst M (1995) Charge-carrier kinetics in semiconductors by microwave conductivity measurements. J Appl Phys 78:1763–1775

145. Kunst M, Abdallah O, Wünsch F (2001) The characterization of silicon nitride films by contactless transient photoconductivity measurements. Thin Solid Films 383:61–64

146. Kunst M, Beck G (1986) The study of charge carrier kinetics in semiconductors by microwave conductivity measurements. J Appl Phys 60:3558–3566

147. Schlichthörl G, Tributsch H (1992) Microwave photoelectrochemistry. Electrochim Acta 37:919–931

148. Schlichthörl G (1992) Untersuchung der Ladungsträgerkinetik in photoelektrochemischen Systemen mit lichtinduzierter Mikrowellenreflexion. PhD thesis, Freie Universität Berlin

149. Kunst M, Abdallah O, Wünsch F (Apr. 2002) Passivation of silicon by silicon nitride films. Sol Energy Mater Sol Cells 72:335–341

150. Sinton RA (2006) WCT-120 Photoconductance lifetime tester and optional Suns-VOC stage, 1132 Green Circle Boulder, CO 80305 USA

151. Sinton RA, Cuevas A (1996) Contactless determination of current–voltage characteristics and minority carrier lifetimes in semiconductors from quasi-steady-state photoconductance data. Appl Phys Lett 69:2510–2512

152. Basore PA, Hansen BR (1990) Microwave-detected photoconductance decay. In: Proceedings of the 21st IEEE photovoltaic specialists conference, Orlando, USA, pp 374–379

153. Sinton RA, Cuevas A (1996) Contactless determination of current–voltage characteristics and minority carrier lifetimes in semiconductors from quasi-steady-state photoconductance data. Appl Phys Lett 69:2510–2512

154. Sinton RA, Cuevas A, Stuckings M (1996) Quasi-steady-state photoconductance, a new method for solar cell material and device characterization. In: Proceedings of the 25th IEEE photovoltaic specialists conference (IEEE-PVSC-25), Washington, USA, pp 457–460

155. Nagel H, Berge C, Aberle AG (1999) Generalized analysis of quasi-steady-state and quasi-transient measurements of carrier lifetimes in semiconductors. J Appl Phys 86:6218–6221

156. Basore PA, Clugston DA (1997) PC1D version 5: 32-bit solar cell modeling on personal computers. In: Proceedings of the 26th IEEE photovoltaic specialists conference, Anaheim, USA, pp 207–210

157. Beck G, Kunst M (Feb. 1986) Contactless scanner for photoactive materials using laser-induced microwave absorption. Rev Sci Instrum 57:197–201

158. Perrin J (1991) Plasma and surface reactions during a-Si:H film growth. J Non-Cryst Solids 137–138:639–644

159. Collins R (1989) Dielectric functions of thin interface layers in a-Si:H-based device structures by spectroscopic ellipsometry. J Non-Cryst Solids 114:160–162

160. Layadi N, Roca i Cabarrocas P, Huc J, Parey J, Drévillon B (1994) In-situ eximer laser induced crystallization of hydrogenated amorphous silicon thin films. Solid State Phenom 37–38:281–286

161. Citarella G, Fahrner W, Neitzert H, Wünsch F, Kunst M (2006) Numerical simulation of time resolved charge transport in semiconductor structures for electronic devices. J Comput Electron 5:211–215

162. Zhao J, Green MA, Wang A (2002) High-efficiency optical emission, detection, and coupling using silicon diodes. J Appl Phys 92:2977–2979

163. Fuyuki T, Kondo H, Yamazaki T, Takahashi Y, Uraoka Y (2005) Photographic surveying of minority carrier diffusion length in polycrystalline silicon solar cells by electro luminescence. Appl Phys Lett 86:262108–3

164. Hoang T, LeMinh P, Holleman J, Schmitz J (2006) The effect of dislocation loops on the light emission of silicon LEDs. Electron Device Lett 27:105–107

165. Dekorsy T, Sun J, Skorupa W, Schmidt B, Helm M (2004) Light-emitting silicon pn diodes. Appl Phys A: Mater Sci Process 78:471–475

166. Tardon S, Roesch M, Brueggemann R, Unhold T, Bauer GH (2004) Photoluminescence studies of a-Si:H/c-Si-heterojunction solar cells. J Non-Cryst Solids 338–340:444–447

167. Bresler MS, Gusev OB, Terukov EI, Fuhs W, Froitzheim A, Gudovskikh AS, Kleider JP, Weiser G (2004) Electroluminescence from amorphous-crystalline silicon heterostructures. J Non-Cryst Solids 338–340:440–443

168. Ferrara M (2009) Electroluminescence of a-Si:H/c-Si heterojunction solar cells after high energy irradiation. PhD thesis, University of Hagen

169. Neitzert HC, Ferrara M, Fahrner W, Scherff M, Klaver A, van Swaaij R (2009) Photocurrent enhancement induced by interface modification due to low dose electron irradiation of amorphous/crystalline silicon heterojunctions. In: Proceedings of the international conference of the physics of semiconductors, Rio de Janeiro, 2008, AIP conference proceedings, vol 1199. pp 17, 18

170. Tompkins HG, McGahan WA (1999) Spectroscopic ellipsometry and reflectometry: a user's guide. Wiley, New York

171. Woolam JA (1997) Guiding to using WVASE32, software for VASE ellipsometers. J. A. Woolam Co., Inc., Lincoln

172. Jellison GEJ, Modine FA (1996) Parameterization of the optical functions of amorphous materials in the interband region. Appl Phys Lett 69:371–373

173. Ferraro JR, Nakamoto K (1994) Introductory Raman spectroscopy. Academic Press, New York

174. Suetaka W (1995) Surface infrared and Raman spectroscopy. Plenum Press, New York

175. Turrell G, Corset J (1996) Raman microscopy—developments and applications. Academic Press, New York

176. Schrader B (1995) Infrared and Raman spectroscopy. VCH Verlagsgesellschaft, Weinheim

177. Laserna JJ (1996) Modern techniques in Raman spectroscopy. Wiley, New York

178. Schroder DK (1998) Semiconductor material and device characterization. Wiley, New York

179. Y Huang (2006) Hydrogen diffusion and Hydrogen related electronic defects in Hydrogen plasma treated and subsequently annealed crystalline Silicon. PhD thesis, University of Hagen

180. Voicu C (2003) Entwicklung und Aufbau einer Steuerung für einen Solarsimulator. Master thesis, FernUniversitaet Hagen

181. Honsberg C, Bowden S (2007) Photovoltaics CDROM. University of Delaware (Solar Hydrogen IGERT)

182. Nelson J (2003) Physics of solar cells. Imperial College Press, London

183. Scheer R, Lewerenz HJ (1995) Diffusion length measurements on n-CuInS$_2$ crystals by evaluation of electron-beam-induced current profiles in edge-scan and planar configurations. J Appl Phys 77:2006–2009

184. Corkish R, Puzzer T, Sproul AB, Luke KL (1998) Quantitative interpretation of electron-beam-induced current grain boundary contrast profiles with application to silicon. J Appl Phys 84:5473–5481

185. Froitzheim A, Stangl R, Elstner L, Schmidt M, Fuhs W (2002) Interface recombination in amorphous/crystalline silicon solar cells, a simulation study. In: Proceedings of the 29th IEEE photovoltaic specialists conference, New Orleans, pp 1238–1241

186. Caputo D, Forghieri U, Palma F (1997) Low-temperature admittance measurement in amorphous silicon. J Appl Phys 82:733–741

187. Roesch M, Brueggemann R, Bauer GH (1998) Influence of interface defects on current voltage characteristics of amorphous silicon/crystalline silicon heterojunction solar cells. In: Proceedings of the 2nd world conference for photovoltaic solar energy conversion, Ispra, Italy, p 964

188. Farrokh Baroughi M, Sivoththaman S (2006) Modeling of grain boundary effects in amorphous/multicrystalline silicon heterojunction solar cells and photodiodes. Semicond Sci Technol 21:979–986

189. Diouf D, Kleider J, Desrues T, Ribeyron P (2010) 2D simulations of interdigitated back contact heterojunctions solar cells based on n-type crystalline silicon. Phys Status Solidi C 7:1033–1036

190. Lu M, Bowden S, Das U, Birkmire R (2007) Interdigitated back contact silicon heterojunction solar cell and the effect of front surface passivation. Appl Phys Lett 91:063507

191. Hernándes-Como N, Morales-Acevedo A (2010) Simulation of hetero-junction silicon solar cells with AMPS-1D. Sol Energy Mater Solar Cells 94:62–67

192. Niemegeers A, Gillis S, Brugelman M (1998) A user program for realistic simulation of polycrystalline heterojunction solar cells: SCAPS-1D. In: Proceedings of 2nd world PVSEC, Vienna, Austria, pp 672–675

193. Pieters BE, Krc J, Zeman M (2006) Advanced numerical simulation tool for solar cells—ASA5. In: Proceedings of the IEEE 4th world conference on photovoltaic energy conversion, pp 1513–1516

194. Neitzert HC, Spinillo P, Bellone S, Licciardo GD, Tucci M, Roca F, Gialanella L, Romano M (2004) Investigation of the damage as induced by 1.7 MeV protons in an amorphous/crystalline silicon heterojunction solar cell. Sol Energy Mater Sol Cells 83:435–446

195. Ziegler JF, Biersack JP, Littmark U (1985) The stopping and range of ions in solids. Pergamon Press, Oxford

196. Borchert D (1998) Technologie von Silizium-Heterosolarzellen mit amorphem und mikrokristallinem Emitter. PhD thesis, Hagen, Germany

197. Olibet S, Vallat-Sauvin E, Ballif C (2006) Effect of light-induced degradation on passivating properties of a-Si:H layers deposited on crystalline si. In: Proceedings of the 21st European photovoltaic conference, Dresden

198. Scherff M, Drzymalla R, Goesse R, Fahrner WR, Ferrara M, Neitzert HC, Opitz-Coutureau J, Denker A (2006) Proton damage in amorphous silicon/crystalline silicon heterojunction solar cells—measurement and simulation. J Electrochem Soc 153:G1117–G1121

199. Braeunig D (1998) Wirkung hochenergetischer Strahlung auf Halbleiterbauelemente, Springer, Berlin

200. Khan A, Yamaguchi M, Aburaya T (2000) Comparison of the effects of electron and proton irradiation on type-converted silicon space solar cells upon annealing. Semicond Sci Technol 15:403–407

201. Meulenberg A, Treble FC (1973) Damage in silicon solar cells from 2 to 155 MeV protons. In: Proceedings of the 10th IEEE photovoltaic specialist conference, Palo Alto, CA, p 359

202. Ricketts LW (1972) Fundamentals of nuclear hardening of electronic equipment. Wiley-Interscience, New York

203. Arora ND, Chamberlain SG, Roulston DJ (1980) Diffusion length determination in p-n junction diodes and solar cells. Appl Phys Lett 37:325–327

204. Basore P (1993) Extended spectral analysis of internal quantum efficiency. In: Proceedings of 23th IEEE photovoltaic specialist conference, New York, pp 147–152

205. Tada HY, Carter JR, Anspaugh BE, Downing RG (1982) Solar cell radiation handbook. NASA, JPL, Pasadena

206. Anspaugh BE, Downing RG (1981) Damage coefficients and thermal annealing of irradiated silicon and GaAs solar cells. In: Proceedings of 15th IEEE photovoltaic specialist conference, Kissimee, FL, pp 499–505

207. 2005 http://rd49.web.cern.ch/RD49/RD49Docs/giustino/Chapter7.pdf

208. Schmidt M, Schoepke A, Korte L, Milch O, Fuhs W (2004) Density distribution of gap states in extremely thin a-Si:H layers on crystalline silicon wafer. J Non-Cryst Solids 338–344:211–214

209. Neitzert HC, Ferrara M, Licciardo GD, Ma Y, Fahrner W, Bobeico E, Delli Veneri P, Mercaldo L, Gialanella L, Romano M, Limata B, Di Bartolomeo A, Ravotti F, Glaser M (2005) Modification of amorphous and microcrystalline silicon film properties after

irradation with MeV and GeV protons. In: Proceedings of 20th EU European photovoltaic solar energy conference, Barcelona, Spain, pp 1627–1630

210. Kinoshita T, Fujishima D, Yano A, Ogane A, Tohoda S, Matsuyama K, Nakamura Y, Tokuoka N, Kanno H, Sakata H, Taguchi M, Maruyama E (2011) The approaches for high efficiency HIT^{TM} solar cell with very thin (<100 μm) silicon wafer over 23%. In: Proceedings of the 26th EU PVSEC, Hamburg, Germany, pp 871–874

211. Bätzner D, Andrault Y, Andreetta L, Büchel A, Frammelsberger W, Guerin C, Holm N, Lachenal D, Meixenberger J, Papet P, Rau B, Strahm B, Wahli, Wünsch F (2011) Characterisation of over 21% efficient silicon heterojunction cells developed at ROTH & RAU Switzerland. In: Proceedings of the 26th EU PVSEC, Hamburg, Germany, pp 1073–1075

212. Choi J-H, Kim S-K, Lee J-C, Park H, Lee W-J, Cho E-C (2011) Advanced module fabrication of silicon heterojunction solar cells using anisotropic conductive film method. In: Proceedings of the 26th EU PVSEC, Hamburg, Germany, pp 3302–3304

213. Hernandez JL, Yoshikawa K, Feltrin A, Menou N, Valckx N, Van Assche E, Poortmans J, Adachi D, Yoshimi M, Uto T, Uzu H, Kuchiyama T, Allebe C, Nakanishi N, Terashita T, Fujimoto T, Koizumi G, Yamamoto K (2011) High efficiency silver-free heterojunction silicon solar cell. In: Proceedings of the 21st PVSEC, Fukuoka, Japan (to be published)

214. Muñoz D, Desrues T, Ozanne AS, Nguyen N, de Vecchi S, Souche F, Martin de Nicolàs S, Denis C, Ribeyron PJ (2011) Progress on high efficiency standard and interdigitated back contact silicon heterojunction solar cells. In: Proceedings of the 26th EU PVSEC, Hamburg, Germany, pp 861–864

215. Descoeudres A, Barraud L, De Wolf S, Strahm B, Lachenal D, Guérin C, Holman ZC, Zicarelli F, Demaurex B, Seif J, Holovsky J, Ballif C (2011) Improved amorphous/crystalline silicon interface passivation by hydrogen plasma treatment. Appl Phys Lett 99:123506

216. Fahrner WR (1991) Private communication

217. Wang Q, Page MR, Iwaniczko E, Xu YQ, Roybal L, Bauer R, To B, Yuan HC, Duda A, Yan YF (2010) Efficient heterojunction solar cells on p-type crystal silicon wafers. Appl Phys Lett 96:013507

218. Wang Q, Page MR, Iwaniczko E, Xu YQ, Roybal L, Bauer R, To B, Yuan HC, Duda A, Yan YF (2010) Efficient heterojunction solar cells on p-type crystal silicon wafers. Appl Phys Lett 96:013507

219. Irikawa J, Miyajima S, Watahiki T, Konagai M (2011) High efficiency hydrogenated Nan[219]ocrystalline cubic silicon carbide/crystalline silicon heterojunction solar cells using an optimized buffer layer. Appl Phys Expr 4:092301

220. Jiang X, Yin W, Li T, Yao K ITO/Si heterojunction solar cells. Acta Energiae Solaris Sinica 3(2):219–221

221. Jiang X, Yin W, Li T SnO$_2$/n-Si heterojunction solar cells. Acta Energiae Solaris Sinica 3(4):457–459

222. Liu Z, Wang D (1987) SnO$_2$/SiO$_x$/poly-Si heterojunction solar cells. J Yunnan Norm Univ (Nat Sci Ed) 3:49–57

223. Zenyu J, Yuhua D, Yu Z, Yuqin Z, Fengzhen L, Meifang Z (2009) Effect of chemical polish etching and post annealing on the performance of the heterojunction solar cells. J Semicond 30:084010

224. Zhou Y, Wu C, Jiang Z, Zhu M (2010) Proceeding of the 11th China solar photovoltaic conference and exhibition, Nanjing, pp 340–343

225. Chen WJ, Wang W et al (2011) Recipe for the Si wafer texturing for large size pyramids. Private seminar, Nov 2011

226. Wang Y, Huang H, Zhou L et al (2012) Private group seminar, Jan 2012

227. Zhou C, Wang W, Zhao L, Li H, Diao H, Cao X (2010) Preparation and characterization of homogeneity and fine pyramids on the texturized single silicon crystal. Acta Phys Sinica 59:5777–5783

228. Zhou Y, Zhang Y, Zhu M, Liu F (2010) Study on sodium phophate texturization in Si-film/
     c-Si heterojunction solar cells. In: Proceedings of the 11th China solar photovoltaic
     conference and exhibition, Nanjing, pp 355–358
229. Zhao L, Zhou C, Li H, Diao H, Wang W (2010) Characterization on the passivation stability
     of HF aqueous solution treated silicon surfaces for HIT solar cell application by the effective
     minority carrier lifetime measurement. Chin J Phys 48:392–399
230. Wang S (2010) Study on silicon wafer processing and transparent conductive multilayers of
     HIT solar cell. Master thesis, Jiangnan University
231. Zhang D, Gong H, Huang H, Zhou L, Fahrner W et al (2011) Private group seminar, Aug
     2011
232. Li J (2011) Study on the passivation for the rear surface of the HIT cells. Bachelor thesis,
     Nanchang University
233. Yuan L (2011) Depositing the SiOx films by CVD method. Bachelor thesis, Nanchang
     University
234. Zhao L, Diao H, Zeng X, Zhou C, Li H, Wang W (2010) Comparative study of the surface
     passivation on crystalline silicon by silicon thin films with different structures. Phys B
     405:61–64
235. Chen X, Wang S, Wang Y, Zhang C, Shi Z (2010) The study of surface passivation of HIT
     solar cells. In: Proceedings of the 11th China solar photovoltaic conference and exhibition,
     Nanjing, pp 325–328
236. Zhou BQ, Liu FZ, Zhang QF, Xu Y, Zhou YQ, Liu JL, Zhu MF (2006) Fabrication of mc-
     Si:H(p)/c-Si(n) heterojunction solar cells in microcrystalline emitters. Chin Phys Lett
     23:1638–1640
237. Zhang Q, Zhu M, Liu F, Zhou Y (2007) High-efficiency n-nc-Si:H/p-c-Si heterojunction
     solar cell. Chin J Semicond 28:96–99
238. Wang M (20078) Study of amorphous crystalline silicon film and HIT solar cell depositing.
     Master thesis, Hebei University of Technology
239. Lu L (2010) Research on the preparation and properties of ZnO:Al and â-FeSi2 thin films
     for solar cells. PhD thesis, Nanjing University of Aeronautics and Astronautics
240. Chen X, Xue J, Zhang D, Sun J, Ren H, Zhao Y, Geng X (2007) Effect of substrate
     temperature on the ZnO thin films as TCO in solar cells grown by MOCVD technique. Acta
     Phys Sinica 56:1567
241. Hao S, Wu J, Fan L, Huang Y, Lin J, Wie Y (2004) The influence of acid treatment of TiO2
     porous film electrode on photoelectric performance of dye-sensitized solar cell. Sol Energy
     76:745–750
242. Zhao L, Zhou Z, Peng H, Cui R (2005) Indium tin oxide thin films by bias magnetron rf
     sputtering for heterojunction solar cells application. Appl Surf Sci 252:385–392
243. Liu X (2007) Studies of indium Tin Oxide thin films on high efficiency HIT solar cells.
     Master thesis, Hebei University of Technology
244. Chen X (2009) Studies of the passivation of HIT solar cells and ZnO transparent conductive
     oxide. Master thesis, Jiangnan University
245. Wang TH, Iwaniczko E, Page MR, Wang Q, Xu Y, Yan Y, Levi D, Roybal L, Bauer R,
     Branz HM (2006) High-efficiency silicon heterojunction solar cells by HWCVD. In:
     Proceedings of the 2006 IEEE 4th world conference on photovoltaic energy conversion,
     pp 1439–1442
246. Shen H, Huang H (2012) Private discussion, Jan 2012
247. Chunbo W, Yuqin Z, Guorong L, Fengzhen L (2011) Influence of the initial transient state
     of plasma and hydrogen pre-treatment on the interface properties of a silicon heterojunction
     fabricated by PECVD. J Semicond 32:096001
248. Zhang Q, Zhu M, Liu F, Liu J, Xu Y (2006) Study of n-type nc-Si:H films and
     heterojunction solar cells by HWCVD. Acta Energiae Solaris Sinica 27:691–694

249. Zhang Q, Zhu M, Liu F (2007) The optimization of interfacial properties of nc-Si:H/c-Si solar cells in hot-wire chemical vapor deposition process. J Mater Sci: Mater Electron 18:S33–S36

250. Hiroyuki F, Kondo M (2007) Impact of epitaxial growth at the heterointerface of a-Si:H/c-Si solar cells. Appl Phys Lett 90:013503

251. Liu F, Cui J, Zhang Q, Zhu M, Zhou Y (2008) Dark I-V characteristics and carrier transport mechanism in nano-crystalline silicon thin film/crystalline silicon hetero-junction solar cell. J Semicond 29:549–553

252. Lin H, Che B (1997) Analysis of technologies in manufacture of a-Si/c-Si heterojunction solar cell. J China Univ Sci Technol 27:132–136

253. Lin H (1999) Numerical analysis of μc-Si/c-Si p-n heterojunction solar cells inserted with a-Si:H thin film at thermodynamic equilibrium. Semicond Technol 24(6):15–19

254. Zhao L, Zhou C, Li H, Diao H, Wang W (2008) Optimizing polymorphous silicon back surface field of a-Si(n)/c-Si(p) heterojunction solar cells by simulation. Acta Physica Sinica 57:3212–3218

255. Zhong C (2011) Theoretical study on interface characteristics of a-Si:H/c-Si heterojunctions. PhD thesis, South China University of Technology

256. Peng G, Wang X, Tang P, Zhou W, Yu J (2007) Simulation and optimization of a-Si:H/c-Si hetero-junction solar cells. In: Proceedings of ISES solar world congress 2007. Solar energy and human settlement, vol 3. pp 1159–1163

257. Lin H, Duan K, Ma L (2002) Analysis of the design for a-Si:H thin film in a-Si/c-Si heterojunction solar cells. J Optoelectron Laser 13:460–464

258. Lin H, Ma L (2003) Analysis of the design for a-Si/c-Si heterojunction structure solar cell. Res Prog SSE 23:470–475

259. Lin H, Duan K, Ma L (2001) Analysis of the design for a-SiC:H thin films in a-SiC/c-Si heterojunction solar cells. Semicond Technol 26:70–74

260. Lin H, Duan K, Ma L (2002) Analysis of design for a-SiC/c-Si heterojunction solar cells. Chin J Semicond 23:492–498

261. Zhao L, Zhou C, Li H, Diao H, Wang W (2008) Optimizing polymorphous silicon back surface field of a-Si(n)/c-Si(p) heterojunction solar cells by simulation. Acta Phys Sinica 57:3212–3218

262. Zhao L, Zhou CL, Li HL, Diao HW, Wang WJ (2008) Design optimization of bifacial HIT solar cells on p-type silicon substrates by simulation. Sol Energy Mater Sol Cells 92:673–681

263. Zhao L, Zhou CL, Li HL, Diao HW, Wang WJ (2008) Role of the work function of transparent conductive oxide on the performance of amorphous/crystalline silicon heterojunction solar cells studied by computer simulation. Phys Status Solidi A 205:1215–1221

264. Li L, Zhou B, Chen X, Han B, Hao L (2009) Back surface field of μc-Si(n)/c-Si(p) heterojunction solar cells by simulation and optimization. Inf Recording Mater 10:18–21

265. Ren B, Zhang Y, Guo B, Zhang B, Li H, Wang W, Zhao L (2007) Computer simulation of p-a-Si:H/n-c-Si heterojunction solar cells. In: Proceedings of ISES solar world congress 2007. Solar energy and human settlement, vol 3. pp 1239–1242

266. Zhong C, Geng KW, Yao R (2010) S-shaped J-V characteristic of a-Si:H/c-Si heterojunction solar cell. Acta Phys Sinica 59:6538–6544